NATURAL DISASTER RESEARCH, PREDICTION AND MITIGATION

HURRICANE SANDY DISASTER RELIEF

SELECT ASSESSMENTS

NATURAL DISASTER RESEARCH, PREDICTION AND MITIGATION

Additional books in this series can be found on Nova's website under the Series tab.

Additional e-books in this series can be found on Nova's website under the e-book tab.

NATURAL DISASTER RESEARCH, PREDICTION AND MITIGATION

HURRICANE SANDY DISASTER RELIEF

SELECT ASSESSMENTS

EDWARD S. RAMIREZ
EDITOR

nova publishers
New York

Additional color graphics may be available in the e-book version of this book.

Library of Congress Cataloging-in-Publication Data

ISBN: 978-1-63482-174-2

Published by Nova Science Publishers, Inc. † New York

CONTENTS

PREFACE

This book discusses the extent to which FEMA implemented controls to help prevent IHP payments that are at risk of being improper or potentially fraudulent; and challenges FEMA and states faced obtaining information to help prevent IHP payments from duplicating or overlapping with other sources in its response to Hurricane Sandy. This book also discusses the the timeliness of SBA's disaster assistance to small businesses; the loan approval rates for small businesses and reasons for decline for Hurricane Sandy and previous disasters; the extent to which SBA has implemented programs mandated by the Small Business Disaster Response and Loan Improvements Act of 2008; the progress DOT has made allocating, obligating, and disbursing DRAA surface transportation funds; how FTA's new Public Transportation Emergency Relief program compares to FEMA's and FHWA's emergency relief programs; and the extent to which FTA and FEMA have implemented their memorandum of agreement to coordinate their roles and responsibilities when providing assistance to transit agencies.

Chapter 1 - Hurricane Sandy struck the United States in October 2012, causing an estimated $65 billion in damages. FEMA provides assistance to survivors through IHP and other programs. Part of its mission is to provide assistance quickly, but GAO previously identified weaknesses in FEMA's ability to do so while protecting government resources. Moreover, GAO's 2006 reports on Hurricane Katrina and Rita showed that FEMA did not consistently validate the identity of applicants or inspect damaged properties.

GAO was asked to review the internal controls FEMA's IHP used in response to the storm. This report discusses (1) the extent to which FEMA implemented controls to help prevent IHP payments that are at risk of being improper or potentially fraudulent and (2) challenges FEMA and states faced obtaining information to help prevent IHP payments from duplicating or overlapping with other sources in its response to Hurricane Sandy.

GAO reviewed FEMA's records for assistance provided to the five states that received IHP following the storm, assessed FEMA's controls to prevent individuals from fraudulently receiving disaster assistance, and interviewed FEMA officials and selected officials from states impacted by the storm.

Chapter 2 - On October 29, 2012, Hurricane Sandy made landfall, causing an estimated $65 billion in damage. SBA administers the Disaster Loan Program, which provides physical disaster loans (used to rebuild or replace damaged property) and economic injury disaster loans (used for working capital until normal operations resume) to help businesses and individual homeowners recover from disasters. In the aftermath of Hurricane Sandy, Congress passed the Disaster Relief Appropriations Act of 2013, which appropriated $779 million to SBA for disaster loans and administrative expenses.

GAO was asked to review SBA's assistance to small businesses following Hurricane Sandy. This report examines (1) the timeliness of SBA's disaster assistance to small businesses; (2) the loan approval rates for small businesses and reasons for decline for Hurricane Sandy and previous disasters; and (3) the extent to which SBA has implemented programs mandated by the Small Business Disaster Response and Loan Improvements Act of 2008. GAO analyzed SBA data on application processing; reviewed documentation related to SBA's planning, relevant legislation, and regulations; and interviewed SBA officials.

Chapter 3 - In October 2012, Hurricane Sandy devastated portions of the Mid-Atlantic coast causing severe damage to transit facilities and infrastructure and disrupting mobility in the New York metropolitan region. In January 2013, the President signed the DRAA, which provided approximately $50.5 billion in federal aid for expenses related to Hurricane Sandy. GAO was asked to examine DRAA emergency relief assistance for transportation.

This report addresses (1) the progress DOT has made allocating, obligating, and disbursing DRAA surface transportation funds, (2) how FTA's new Public Transportation Emergency Relief program compares to FEMA's and FHWA's emergency relief programs, and (3) the extent to which FTA and FEMA have implemented their memorandum of agreement to coordinate their roles and responsibilities when providing assistance to transit agencies. GAO analyzed relevant laws, regulations, and agency documentation, and interviewed DOT, FEMA, and New Jersey and New York area transit officials.

In: Hurricane Sandy Disaster Relief
Editor: Edward S. Ramirez
ISBN: 978-1-63482-174-2

Chapter 1

HURRICANE SANDY: FEMA HAS IMPROVED DISASTER AID VERIFICATION BUT COULD ACT TO FURTHER LIMIT IMPROPER ASSISTANCE[*]

United States Government Accountability Office

WHY GAO DID THIS STUDY

Hurricane Sandy struck the United States in October 2012, causing an estimated $65 billion in damages. FEMA provides assistance to survivors through IHP and other programs. Part of its mission is to provide assistance quickly, but GAO previously identified weaknesses in FEMA's ability to do so while protecting government resources. Moreover, GAO's 2006 reports on Hurricane Katrina and Rita showed that FEMA did not consistently validate the identity of applicants or inspect damaged properties.

GAO was asked to review the internal controls FEMA's IHP used in response to the storm. This report discusses (1) the extent to which FEMA implemented controls to help prevent IHP payments that are at risk of being improper or potentially fraudulent and (2) challenges FEMA and states faced

[*] This is an edited, reformatted and augmented version of a United States Government Accountability Office publication, No. GAO-15-15, dated December 2014.

obtaining information to help prevent IHP payments from duplicating or overlapping with other sources in its response to Hurricane Sandy.

GAO reviewed FEMA's records for assistance provided to the five states that received IHP following the storm, assessed FEMA's controls to prevent individuals from fraudulently receiving disaster assistance, and interviewed FEMA officials and selected officials from states impacted by the storm.

WHAT GAO RECOMMENDS

GAO recommends, among other things, that FEMA collaborate with SSA to obtain additional data, collect data to detect duplicative assistance, and implement an approach to verify whether recipients have private insurance. FEMA concurred with the recommendations.

WHAT GAO FOUND

By implementing new controls since the mid-2000s, the Federal Emergency Management Agency (FEMA) improved its ability to detect improper and potentially fraudulent payments, but GAO identified continued weaknesses in the agency's validation of Social Security numbers, among other things. As of August 2014, FEMA stated that it had provided over $1.4 billion in Hurricane Sandy assistance through its Individuals and Households Program (IHP)—which provides financial awards for home repairs, rental assistance, and other needs— to almost 183,000 survivors. GAO identified $39 million, or 2.7 percent, that was at risk of being improper or fraudulent, compared to 10–22 percent of similar assistance provided for Hurricanes Katrina and Rita. GAO identified payments as at-risk if they had characteristics indicating, for example, that ineligible recipients or duplication of assistance could be involved. However, it is not possible to determine whether these payments were definitively improper or fraudulent without inspecting each payment. FEMA officials reviewed GAO's findings and stated that at least $6.1 million of the $39 million were not improper or fraudulent, but GAO could not independently confirm their conclusions for each payment. In February 2006, FEMA began using a tool to validate the identity of applicants during registration. FEMA also hired contractors to inspect damaged homes to verify the identity and residency of applicants and that reported damage was a result of Hurricane Sandy. However, in this review, GAO found 2,610 recipients with potentially invalid identifying information who received $21 million of the $39 million GAO calculated as potentially improper or fraudulent. GAO's analysis included data from the Social Security Administration (SSA) that FEMA does not use, such as SSA's most-complete death records. Collaborating with SSA prior to providing assistance could give FEMA additional information to further reduce its risk of assisting ineligible applicants.

FEMA and state governments faced challenges in obtaining the data necessary to help prevent duplicative payments from overlapping sources. For example, FEMA was unable to identify potentially duplicative rental-assistance payments to recipients of its Sheltering and Temporary Essential Power pilot program, in part because it did not request the necessary data from states at the program's outset. FEMA recently took steps to make data sharing among programs easier, including initiating a committee to explore ways to maintain and share relevant data needed to evaluate and help prevent potentially duplicative assistance in the future. In addition, FEMA relies on self-reported data from applicants regarding private home insurance—a factor the agency uses in determining benefits, as federal law prohibits FEMA from providing assistance for damage covered by private insurance. By examining data from entities that provide federally backed mortgages, GAO identified 534 individuals receiving over $2.3 million in home repair and personal property assistance who said they did not have private insurance but had mortgages that require such insurance. FEMA reviewed 55 cases and stated that 32 were likely appropriate because the assistance was for damage not covered by private insurance. However, the risk remains that some individuals may have received assistance from FEMA for ineligible expenses. Assessing a variety of methods to verify self-reported data would provide FEMA greater assurance that it has a cost-effective means of obtaining sufficiently accurate and reliable information.

ABBREVIATIONS

DHS	Department of Homeland Security
DHS OIG	Department of Homeland Security Office of the Inspector General
EVS	Enumeration Verification System
Fannie Mae	Federal National Mortgage Association

FEMA	Federal Emergency Management Agency
FHA	Federal Housing Administration
Freddie Mac	Federal Home Loan Mortgage Corporation
IHP	Individuals and Households Program
NEMIS	National Emergency Management Information System
NFIP	National Flood Insurance Program
ONA	Other Needs Assistance
PKEMRA	Post-Katrina Emergency Management Reform Act of 2006
SSA	Social Security Administration
SSN	Social Security number
Stafford Act	Robert T Stafford Disaster Relief and Emergency Assistance Act
STEP	Sheltering and Temporary Essential Power
TSA	Transitional Shelter Assistance

* * *

December 12, 2014
Congressional Requesters

In its mission to reduce the loss of life and property after disasters, the Federal Emergency Management Agency (FEMA) must quickly assist those in need while also protecting government resources from fraud, waste, and abuse. Both objectives are critical to the agency's success and strong governance, but achieving them simultaneously can be challenging. If FEMA provides assistance too quickly, abuse may go undetected; conversely, if FEMA spends too much time assessing applicants' eligibility, the applicants may be deprived of needed assistance and services. Our work has consistently shown that reliable and effective controls are vital in helping achieve an appropriate balance.[1]

In 2005, FEMA provided $7 billion in financial assistance to survivors of Hurricanes Katrina and Rita, which caused an estimated $159.4 billion in damages. The response to these disasters exposed several vulnerabilities in FEMA's internal controls. In a series of reports issued from June through December 2006, we identified significant control weaknesses that resulted in an estimated $600 million to $1.4 billion in potentially fraudulent or improper payments.[2] FEMA reported that approximately $60 million of those payments had been recovered as of September 2014.[3] Our 2006 work examining FEMA's response to Hurricanes Katrina and Rita showed that FEMA did not consistently validate the identity of applicants, inspect or confirm the physical existence of damaged residences, or confirm whether potential aid recipients actually owned or occupied a damaged property at the time of the disaster. In June and September 2006, we recommended that FEMA develop processes for validating applicants' identities and primary residences during the application process.[4] In response, FEMA took several steps to strengthen its controls, including establishing identity verification for all applications as of June 2006 and inspections for all residences as of 2008. In a June 2009 report, we reexamined these controls in FEMA's 2008 response to Hurricanes Gustav and Ike and found that they had improved.[5] However, we recommended additional actions in that report, including a recommendation that FEMA conduct random checks to assess the validity of applicants' supporting documentation. Beginning in December 2009, the agency took further steps to improve controls by randomly reviewing and verifying the legitimacy of the supporting documentation that caseworkers accept from applicants under the Individuals and Households Program (IHP), which provides financial awards for home repairs, rental assistance, and other needs. As in the 2005 response to Hurricanes Katrina and Rita and the 2008 response to Hurricanes Gustav and Ike, FEMA provided large-scale posthurricane assistance to survivors after Hurricane Sandy struck the Northeast United States in October 2012 and caused an estimated $65 billion worth of damage. The Disaster Relief Appropriations Act of 2013 included approximately $50 billion in supplemental appropriations for fiscal year 2013 for expenses related to the consequences of Hurricane Sandy.[6] The funds were allocated to a number of agencies, including almost $11.8 billion to FEMA, which used a portion of the funds to provide housing and other assistance to those affected by Hurricane Sandy through a number of programs, including to IHP. According to the data FEMA provided in November 2013, the most current and complete records at the time of our request, 182,782 individuals in five disaster-affected states received assistance through the agency's IHP program.[7]

You asked us to review the internal controls FEMA used in providing assistance through IHP as a part of the agency's response to Hurricane Sandy. This report assesses (1) the extent to which FEMA implemented controls to help prevent IHP payments that are at risk of being improper or potentially fraudulent and (2) what challenges, if any, FEMA and states faced in obtaining and sharing information to help prevent IHP payments from duplicating or overlapping with other sources of assistance in FEMA's response to Hurricane Sandy.[8]

To assess the extent to which FEMA had implemented controls to help prevent payments that are at risk of being improper or potentially fraudulent, we reviewed FEMA's IHP records for applications and assistance provided in response to Hurricane Sandy as of November 2013, as well as FEMA's controls designed to prevent individuals from fraudulently receiving disaster assistance.[9] We submitted IHP applicants' Social Security numbers (SSN) to the Social Security Administration's (SSA) Enumeration Verification System (EVS). The EVS provides information on invalid (never issued) SSNs and instances where there are mismatches between SSN, name, and birth date. In addition, we searched the IHP applicant database for records with duplicative SSNs, addresses, bank account numbers, and phone numbers as well as evidence that personal information of likely deceased individuals on SSA's full death file was used to obtain IHP benefits.[10] We also compared IHP records with prison records of individuals incarcerated in state prisons in Connecticut, New Jersey, and New York to determine whether rental assistance payments were made to prisoners.[11] Finally, we interviewed FEMA officials and reviewed policies related to eligibility for assistance for flood-related damage through IHP and reviewed IHP assistance records to assess the extent to which that guidance was consistently followed.

To examine the challenges, if any, FEMA and states faced in obtaining and sharing information to help prevent duplicative IHP payments from overlapping sources, we reviewed application data of IHP recipients who reported that they did not have private homeowners insurance to determine whether their damaged properties had a federally backed mortgage, which housing and mortgage officials told us require mortgage holders to maintain private insurance. We compared these applications with data from the Federal Housing Administration (FHA), the Federal National Mortgage Association (Fannie Mae), and the Federal Home Loan Mortgage Corporation (Freddie Mac) to determine whether any IHP recipients may have received assistance for damage from Hurricane Sandy that could be covered by private homeowners insurance.

We also interviewed state and federal officials and reviewed guidance about FEMA's Sheltering and Temporary Essential Power (STEP) pilot program to assess FEMA's ability to determine whether an applicant's home had been repaired through the STEP program prior to providing sheltering assistance through the IHP.[12]

We assessed the accuracy and reliability of FEMA's IHP data and data provided by state departments of correction by interviewing agency officials knowledgeable about the data system and by obtaining written responses regarding (1) methods of data collection and quality control reviews, (2) practices and controls over data-entry accuracy, and (3) any limitations of the data.

To assess the reliability of data from FHA, Fannie Mae, and Freddie Mac, we interviewed agency officials knowledgeable about the data system and mortgage loans. In addition, we conducted electronic testing of the data, including checks for missing, out of range, or logically inaccurate data. We determined that the data were sufficiently reliable for the purposes of this report. We assessed FEMA's controls and the challenges FEMA and states faced in obtaining data to prevent overlapping payments by using criteria from FEMA guidance, a review of laws pertaining to disaster relief, including the Post-Katrina Emergency Management Reform Act of 2006 (PKEMRA),[13] and the Robert T. Stafford Disaster Relief and Emergency Assistance Act (Stafford Act), as amended.[14]

We also reviewed best practices from auditing and antifraud organizations, internal control criteria from *Standards for Internal Control in the Federal Government*, and previous GAO work on fraud prevention and information systems controls.[15] We conducted this performance audit from July 2013 to December 2014 in accordance with generally accepted government auditing standards. Those standards require that we plan and perform the audit to obtain sufficient, appropriate evidence to provide a reasonable basis for our findings and conclusions based on our audit objectives.

We believe that the evidence obtained provides a reasonable basis for our findings and conclusions based on our audit objectives. A more detailed explanation of our objectives, scope, and methodology can be found in appendix I.

BACKGROUND

In late October 2012, Hurricane Sandy made landfall in southern New Jersey, with effects felt across 11 states and the District of Columbia, particularly the densely populated New York and New Jersey coasts. The storm caused an estimated $65 billion worth of damage, as more than 23,000 people sought refuge in temporary shelters and more than 8.5 million utility customers lost power. In response, the Disaster Relief Appropriations Act of 2013 was signed into law in January 2013, which included approximately $50 billion in supplemental appropriations for expenses related to the consequences of Hurricane Sandy. The funds were allocated to a number of agencies, including almost $11.8 billion to FEMA, which used a portion of the funds to provide housing and other assistance to those affected by Hurricane Sandy through a number of programs. As of September 30, 2014, the agency had obligated nearly $10.6 billion to Hurricane Sandy recovery, of which over $1.6 billion was expended through IHP assistance.

FEMA Disaster Assistance Programs

FEMA disaster assistance is provided through its IHP, which is part of its Individual Assistance program, as well as through its Public Assistance grant programs. These federal assistance programs were established under the Stafford Act. In addition, the National Flood Insurance Program (NFIP) provides insurance coverage for flood damage to individuals who live in designated flood zones. (See fig. 1 for a summary of the various types of assistance.)

FEMA's IHP program provides financial assistance and direct assistance to eligible individuals and households who, as a direct result of a major disaster or emergency, have uninsured or underinsured necessary expenses and serious needs that cannot be addressed by other means, such as other assistance programs or insurance. For example, IHP provides financial assistance in the form of money to individuals and households for home repairs, rental assistance, personal property, transportation, and funeral and burial costs. FEMA determines whether individuals or households meet eligibility requirements for IHP assistance after they register either online, in person, or over the telephone.[16] Individuals and households may be eligible for assistance if, due to the disaster, they have been displaced from their primary residence or their primary residence has been rendered uninhabitable. Additionally, the primary residence must be located in a presidentially declared disaster area and the applicant must have insufficient or no insurance coverage. Applicants must submit identification information, including name, SSN, and date of birth. Applicants must also provide a physical property address affected by the disaster.

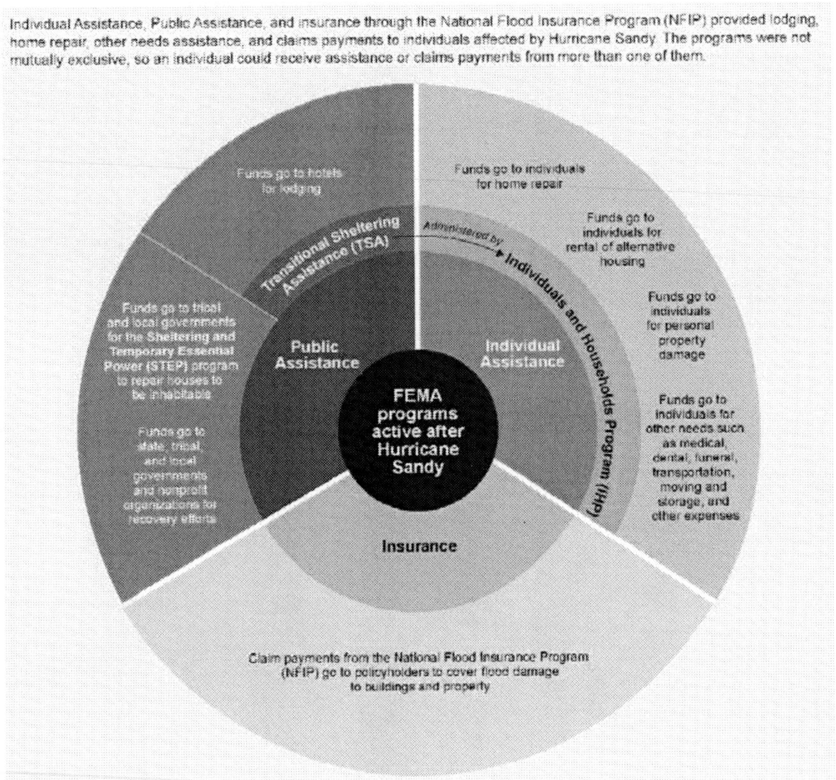

Source: GAO analysis of FEMA data. | GAO-15-15.

Figure 1. Federal Emergency Management Agency (FEMA) Programs Providing Assistance after Hurricane Sandy.

After Hurricane Sandy, individuals or households could receive a maximum of $31,900 in financial assistance through IHP. [17] As of November 2013, FEMA had provided more than $1.4 billion in IHP cash assistance to survivors of Hurricane Sandy.[18] Appendix II shows the number of recipients and total dollar amounts of assistance provided through IHP after Hurricane Sandy, by county.

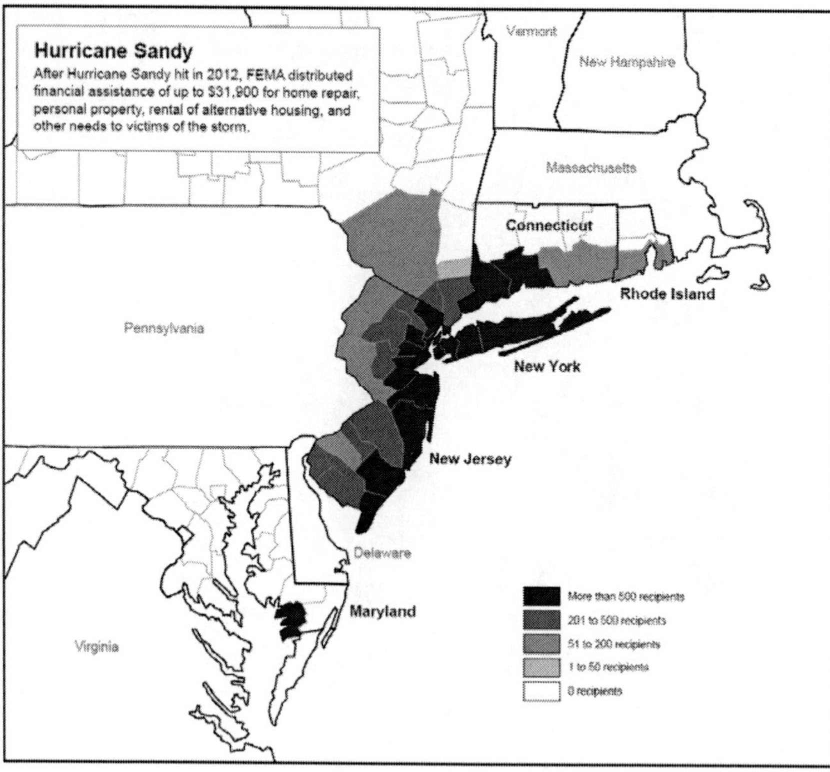

Source: GAO analysis of Federal Emergency Management Agency data (data); Map Resources (map). | GAO-15-15.

Figure 2. Hurricane Sandy Individuals and Households Program (IHP) Recipients by County.

See Appendix II for further information regarding Figure 2.

Table 1 describes the total payments made through the IHP in response to Hurricane Sandy, by state, according to data provided to us by FEMA in November 2013.

FEMA's Public Assistance grant program provides funding to state, tribal, and local governments and nonprofit organizations for recovery efforts after a disaster, including removing debris, implementing emergency protective measures, and repairing or replacing damaged public equipment or facilities. FEMA also uses Public Assistance grants to provide short-term sheltering for disaster survivors who are unable to return to their homes. For instance, Transitional Shelter Assistance (TSA) provides short-term sheltering assistance to disaster survivors who have a continuing need for shelter after congregate shelters, such as those in schools, have closed.[19] Under TSA, disaster survivors may be eligible to stay in hotel or motel lodging for a limited period and have the cost of the room and taxes covered by FEMA. Payments for TSA lodging are made through a contracted vendor directly to participating hotels on behalf of FEMA; disaster survivors are not provided with any cash assistance under this program. While the TSA program is funded under Section 403 of the Stafford Act through Public Assistance, it is administered through FEMA's Individual Assistance program, which is more prepared to collect information on individuals than Public Assistance. FEMA considers applicants' eligibility for TSA as part of the IHP application process.

Table 1. Individuals and Households Program (IHP) Expenditures as of November 2013

State (disaster number)	Number of households receiving assistance	Housing assistance (dollars)	Other needs assistance (dollars)	Total assistance (dollars)
New York (4085)	117,578	$859,928,592	$145,584,045	$1,005,512,637
New Jersey (4086)	61,398	359,074,194	56,748,678	$415,822,872
Connecticut (4087)	2,969	13,976,857	1,107,837	$15,084,694
Maryland (4091)[a]	732	2,147,394	424,167	$2,571,561
Rhode Island (4089)	105	378,749	42,592	$421,341
Total	182,782	$1,235,505,786	$203,907,319	$1,439,413,105

Source: Federal Emergency Management Agency (FEMA). | GAO-15-15

[a]States may choose to participate in the administration of the Other Needs Assistance (ONA) provision of IHP. FEMA administered ONA on behalf of all affected states except for Maryland, which opted to participate in the administration of this benefit.

In addition to IHP and TSA, which were administered by FEMA's Individual Assistance program, FEMA implemented the STEP initiative in the weeks after Hurricane Sandy and reimbursed local governments the cost for providing this assistance to disaster survivors through the Public Assistance program. STEP was a pilot program intended to reduce the demand for other FEMA-provided shelter, such as congregate shelters or TSA, and allow individuals to return to or remain in their homes.[20] Limiting the need for sheltering assistance outside of residents' homes was particularly important because Hurricane Sandy affected an area with high population density, limited space, and high demand for hotel rooms and other potential sheltering facilities. The STEP program shares some commonality with IHP, as it provides basic repairs to disaster-affected residences such as for electricity, heat, hot water, securing broken windows, and covering exterior walls and roofs. However, STEP is intended only to make the homes habitable for the short term and does not provide any assistance beyond basic repairs to homes, whereas IHP provides financial assistance for home repair or replacement, rental expenses, damaged personal property, and other needs. In New York City, the New York City Rapid Repair Program augmented STEP to allow for repairs to multifamily dwellings and more permanent repairs needed to meet building codes. FEMA reported that STEP had provided repairs to approximately 12,500 separate households in New Jersey and New York. FEMA also reported that, as of June 2014, nearly $418 million had been expended through the STEP program to states and localities, which were responsible for establishing contracts with service providers to assist residents in need of repairs. These states and localities later provided FEMA with the records of service, which were maintained at the state or local level, for reimbursement.

Finally, homeowners with mortgages held by federally regulated private lenders on property in participating communities identified by FEMA to be in special flood hazard areas are required to purchase flood insurance under the National Flood Insurance Program (NFIP). Under NFIP, the federal government assumes the liability for the insurance coverage and sets rates and coverage limitations, among other responsibilities, while private insurers sell the policies and administer the claims for a fee determined by FEMA. While certain homeowners are required to purchase flood insurance, any property owner regardless of flood risk may purchase NFIP flood insurance if their community participates in the program. NFIP is administered by FEMA's Federal Insurance and Mitigation Administration on behalf of the federal government, which is responsible for administering programs that provide assistance for mitigating future damage from natural hazards.[21] Homeowners who fail to maintain flood insurance despite being required to do so are usually not eligible for repair, replacement, or restoration assistance from FEMA or other federal agencies for insurable damage related to flooding.

Table 2 summarizes the disaster-recovery programs discussed in this report.

Table 2. Federal Emergency Management Agency (FEMA) Disaster-Recovery Programs and Status of Hurricane Sandy Expenditures as of November 2013

Disaster-recovery program	Type of program	Payment or assistance provided	Number of households receiving assistance or claims payments following Hurricane Sandy	Funds expended or claims paid following Hurricane Sandy (millions of dollars)
Individuals and Households Program (IHP)	Individual Assistance	Up to $31,900 for home repair, rent, and other disaster-related expenses	182,782a	$1,439
Transitional Shelter Assistance (TSA)	Public Assistance	FEMA-paid lodging in local hotels	11,505	106
Sheltering and Temporary Essential Power (STEP)	Public Assistance	Basic home repairs to allow survivors to return quickly	Approximately 12,500	418b
National Flood Insurance Program (NFIP)	Insurance	Insurance claims to policyholders for flood damage	144,293	8,202c

Source: FEMA and GAO analysis of FEMA data. | GAO-15-15

[a]Since FEMA provided the data in November 2013, IHP assistance had increased to $1,454 million for 182,911 recipients as of August 2014.

[b]Data for STEP were provided in September 2014. [c]Data for NFIP data are current as of June 2014.

New Controls Introduced after Hurricanes Katrina and Rita

In responding to disasters and federal legislation since Hurricanes Katrina and Rita, FEMA implemented new controls designed to improve its capacity to verify applicants' eligibility for assistance. We reported in June 2006 that in response to Hurricane Katrina, FEMA made between $600 million to $1.4 billion in improper and potentially fraudulent payments, in part due to insufficient steps by FEMA to verify applicants' identities and residences.[22] Verifying identities is an essential step in preventing fraud, waste, and abuse.[23] After Hurricanes Katrina and Rita, we also identified potential fraud and abuse related to the temporary housing assistance FEMA provided to individuals displaced by the disasters. For example, in the wake of these hurricanes, we reported that some survivors were simultaneously staying in hotels under TSA and receiving cash rental assistance payments.[24] In addition, in March 2007 we reported that FEMA made nearly $20 million in duplicate payments to over 7,000 applicants who improperly claimed the same damaged property for both Hurricanes Katrina and Rita.[25] We made a number of recommendations in our March 2007 report to help minimize duplicate payments, which FEMA has implemented as discussed below.

FEMA instituted a number of changes following Hurricanes Katrina and Rita and implemented new fraud-prevention controls in response to Hurricanes Gustav and Ike in 2008. As we reported in June 2009, FEMA established identity and address verification on all applications, identified duplicate registrations, and increased requirements for home inspections. In addition, FEMA began requiring individuals in need of housing assistance to provide valid registration numbers before checking into FEMA-paid-for hotels. 26

In addition, FEMA took steps to respond to requirements included in the Post-Katrina Emergency Management Reform Act of 2006 (PKEMRA). For example, PKEMRA required FEMA to develop and implement a training program to prevent waste, fraud, and abuse. In response, FEMA now requires all employees to take a fraud-prevention course each calendar year and contractors for FEMA Disaster Housing Inspections provide fraud and abuse avoidance training to the inspectors.[27]

PKEMRA also required that FEMA submit any database or processing system to the Department of Homeland Security Office of the Inspector General (DHS OIG) for a review of internal controls. In response, the DHS OIG has issued a series of reports on FEMA's internal controls. For example, in 2009, a DHS OIG analysis of disaster registration and payment data concluded that FEMA made virtually no payments to applicants with invalid or duplicate SSNs and made far fewer payments to applicants who had provided other duplicate information for the 2008 disasters.[28] In the same report, the DHS OIG reported that IHP payments to duplicate, improper, and potentially fraudulent registrations decreased substantially since Hurricane Katrina, primarily because FEMA required inspection of every applicant's home before it would process an assistance payment.

FEMA Implemented Controls to Prevent Assistance to Ineligible Recipients, but Opportunities Exist to Further Enhance Oversight

The controls that FEMA implemented have improved the agency's ability to prevent improper or potentially fraudulent payments, and the amount of such payments for Hurricane Sandy decreased compared to Hurricanes Katrina and Rita. However, we identified some limitations in the agency's validation of Social Security numbers (SSN), identification of duplicate applications, and oversight of flood damage assistance.

FEMA Improved Controls for Preventing Potentially Improper or Fraudulent IHP Payments prior to Hurricane Sandy

While safeguards were generally not effective after Hurricanes Katrina and Rita, the controls FEMA implemented since then have improved the agency's ability to prevent improper or potentially fraudulent IHP payments. After ensuring that payments in multiple at-risk categories were not counted more than once (as discussed below), we identified about $39 million, or 2.7 percent of the more than $1.4 billion in IHP assistance, that was at risk of being improper or potentially fraudulent after Hurricane Sandy.[29] By comparison, the estimated proportion of improper and potentially fraudulent IHP payments made in response to Hurricanes Katrina and Rita was between $600 million to $1.4 billion, or 10 to 22 percent, of the $6.3 billion of assistance provided. FEMA reports that it is in the process of trying to recover some of the improper payments from Hurricane Sandy and reported that it has requested repayment of over $8.5 million and had recouped about 8 percent of that amount as of October 2014.

As described in more detail through the rest of the report, we defined improper and potentially fraudulent payments as including assistance that was provided to (1) state prisoners, (2) individuals using SSNs that did not match SSA's records or belonged to likely deceased individuals, (3) applicants using the same identifying information as other applicants, and (4) applicants receiving payments for damage that were at risk of being duplicative with damage covered by a private homeowners insurance policy. It also includes payments for flood-related damage for which documentation was not sufficiently transparent to provide assurance that the payments were for eligible damage. In our calculation, however, the $39 million in at-risk dollars is the maximum amount at risk that we identified, and we referred each of these cases to FEMA for further investigation. Based on several examples provided by FEMA in response to our analysis, we would expect case-by-case investigations of these payments to show that some of the payments were proper. Thus, the actual percentage of fraudulent or improper payments could ultimately amount to less than the Office of Management and Budget's benchmark for identifying programs with significant improper payments.[30]

Table 3. Summary of IHP Assistance Identified as At-Risk for Potential Fraud or Improper Payments

Type of risk for payments made	Number of IHP registrations	At-risk assistance (dollars)
Receipt of rental assistance while incarcerated	2	$7,328.00
Social Security number (SSN) was invalid, did not match Social Security Administration's (SSA) records for other identifying information, or SSN belonged to likely deceased individual	2,610	21,892,332.43
Duplicates not flagged by the Federal Emergency Management Agency (FEMA)	203	1,679,836.63
SSNs used by multiple recipients	20	180,725.43
Bank accounts used by multiple recipients	147	1,247,660.29
Addresses where more than 1 applicant received home-repair assistance	128	1,324,189.94
Address and last name used by multiple recipients and cumulatively receiving more than $31,900	418	1,153,336.95
Receipt of funeral assistance through a second registration in the same household	18	71,716.16
Receipt of home-repair assistance and claiming to lack private homeowner's insurance while	534	2,324,748.49

Type of risk for payments made	Number of IHP registrations	At-risk assistance (dollars)
maintaining federally backed mortgage	534	2,324,748.49
Inaccurate flags for National Flood Insurance Program (NFIP) compliance	1,322	10,542,717.72

Source: GAO analysis of FEMA data. | GAO-15-15

Note: Because some of the payments we identified as being at risk appear in more than one category, the sum of the dollar amounts for each category in the table is greater than our total at-risk calculation of $39 million, which counts the payments included in multiple categories only once.

Table 3 lists the categories of at-risk payments we identified, along with the frequency and dollar amount of the payments. Because some of the payments we identified as being at risk appear in more than one category, the sum of the dollar amounts for each category in the table is greater than our total at-risk calculation of $39 million, which counts the payments included in multiple categories only once. Additionally, and as discussed below, FEMA officials stated that at least $6.1 million of the payments we identified as being at risk were deemed appropriate upon the agency's subsequent review of our matching results and related applicant files (see app. III). However, we were not able to independently verify what FEMA officials told us without additional investigation into each payment. Accordingly, while we present what FEMA officials told us about the outcome of their review as context from the agency's perspective, we continue to include these cases in our at-risk calculation.

Two Applicants Incarcerated in State Prisons Obtained Rental Assistance through IHP but May Have Been Eligible for Assistance

The Stafford Act specifies that eligibility for temporary housing assistance is predicated on individuals or households being displaced from their predisaster primary residence or their predisaster primary residence being rendered uninhabitable or inaccessible due to the disaster. FEMA may provide temporary housing assistance in the form of financial assistance to individuals or households to rent alternate housing accommodations. Generally, the eligibility criteria will exclude incarcerated individuals from receiving financial assistance for the time they are in prison, as they do not have a disaster-related need for housing.[31] In a 2006 testimony, however, we described how FEMA paid over $3 million to over 1,000 applicants who used names and SSNs belonging to state and federal prisoners for expedited and housing assistance following Hurricane Katrina.[32] In July 2012, we followed up on our recommendation that FEMA develop a means for preventing prisoners from receiving disaster assistance to which they were ineligible. In response, FEMA reported that it had implemented a control to flag applications using high-risk addresses, including prisons, for further review prior to providing assistance.

To test these controls, we compared data from IHP with data from state prisons in the three most-affected states. 33 Our analysis of FEMA data showed that out of 182,782 recipients, FEMA provided IHP assistance to 2 individuals who were incarcerated for the entire 6 months following Hurricane Sandy. FEMA officials stated that both of the individuals, or adult members of their households, met their procedural requirements for assistance, which includes the owner of a damaged home or another individual over the age of 18 who lived in the home prior to the disaster to be present for the home inspection of the damaged addresses. However, as noted above, while FEMA was able to tell us about the outcome of its review, additional investigation would be required to validate FEMA's conclusions. Thus, we did not independently verify documentation that these individuals were eligible or were present for the inspection.

FEMA's Internal Controls for Identifying Invalid SSNs Did Not Prevent Payments to Some Potentially Ineligible Recipients

In February 2006, FEMA began using a tool created by a contractor to help validate the identity of applicants during the initial review of an application, eliminating any wait otherwise necessary for another agency to validate the applicant's personal information. The validity check cross-references information provided by applicants against multiple databases, including public records, credit reports, and other sources. According to FEMA officials, the validity check uses SSA's Death Master File, which is a publicly available subset of the death records that SSA maintains on deceased SSN-holders.[34] Officials further stated that the validity check compares the applicant's SSN and other identifying information with these various public sources of information to verify that the SSN matches the name of the applicant, does not belong to a deceased individual, and is not associated with more than one person (e.g., the SSN has been used in an instance of identity theft). Officials said that in addition to verifying applicants' SSNs, the validity check provides other checks. For example, it compares damaged property addresses with public records to verify that the applicant lived at the damaged property. It also flags "high-risk" addresses, such as addresses of businesses and prisons.

While FEMA's SSN validity check is designed to identify and flag ineligible applicants, we found that 2,610 of the 182,782 IHP recipients had SSNs that could either not be validated by SSA or appeared on the full death file. These individuals received over $21 million in assistance through the IHP. We identified these 2,610 recipients for potential ineligibility by using multiple sources of SSA data. Specifically, we compared IHP applicants' names, dates of birth, and SSNs to SSA's records using SSA's Enumeration Verification System (EVS). EVS flags SSNs in which the name or date of birth (or both) do not match its records for the SSN, as well as SSNs that have never been issued. We also compared IHP applicants to SSA's full death file to identify individuals who may have used a likely deceased person's identity to receive assistance. As discussed earlier, the full death file lists all SSNs of people for whom SSA has received a record of death, and FEMA uses a subset of the full death file as part of its validity checks.

Of the 2,610 recipients, our analysis showed that 10 recipients used SSNs that had never been issued. These 10 recipients received 14 assistance payments totaling less than $43,000, about 0.004 percent of the payments made to the 182,782 recipients in our review. While this may indicate a limitation in FEMA's validity check, FEMA nevertheless processed fewer payments to applicants using SSNs that had never been issued, suggesting that on the whole FEMA's ability to screen applicants based on invalid SSNs improved. For instance, in our previous work on Hurricanes Katrina and Rita, we estimated that 0.8 percent (or 20,800 payments) of the 2.6 million payments made through the IHP were to applicants who used SSNs that had never been issued.[35]

FEMA reviewed a nongeneralizable sample of 41 of the 2,610 recipients we identified with names or dates of birth that did not match SSA's records and found that all 41 passed its validity check for SSNs. During its review of our analysis, FEMA officials indicated that they used a different third-party vendor's software from the one FEMA uses when it initially receives an application for assistance to verify that the payments were proper. Based on its results, FEMA suggested that a majority of recipients in the sample could have been flagged by EVS because they had changed their last names due to marriage or divorce but had not reported the change to SSA. While this may be case, individuals who legally change their name are to report the change to SSA. Hence, applicants with names that do not match SSA's records are to provide evidence of their legal name prior to receiving assistance from federal programs like IHP. These 41 recipients were not selected randomly; therefore, conclusions regarding the appropriateness of the payments for the 2,610 recipients we identified cannot be made from the sample FEMA reviewed.

We also found that 45 of the 2,610 recipients appeared on the full death file and had an IHP application dated after the reported date of death.[36] These 45 recipients received over $465,000 in assistance. In response to our request for additional information about these cases, FEMA selectively reviewed 36 of the cases and determined that in 26 of these cases either a coapplicant was eligible for assistance because he or she was both an occupant of the damaged property and an heir to the property due to the death of the owner; or there was no indication in the file supporting the situation that the applicant was deceased at the time of the disaster, at inspection, or during calls made by the joint field office.[37] FEMA officials stated that in the remaining 10 cases, 7 were submitted for recoupment review, 2 had payments returned voluntarily, and the last was likely under investigation for fraud.

While some of the information FEMA provided for 26 of the cases may explain why a coapplicant is eligible for an assistance payment made to a deceased applicant, other instances related to the other recipients will require additional follow up by FEMA to determine whether the payments were potentially improper or fraudulent. For instance, when there is no indication in a file that an applicant is deceased, it could be that FEMA's identity check did not detect the use of a deceased person's identity. Furthermore, FEMA allows individuals other than the applicant to meet with an inspector and handle other aspects of the application—a measure often important for applicants who are physically unable to meet with an inspector because, for example, they took shelter with family out of town or are too ill or infirm to travel to the damaged home. However, if FEMA's verification check did not identify a deceased individual, this flexibility in who meets with the inspector can increase the risk of fraud.

As demonstrated by our analysis, FEMA's control to ensure the validity of applicants' SSNs—that the SSN belongs to the applicant and does not belong to a likely deceased individual—could be enhanced if it obtained more-comprehensive sources of information, such as SSA's full death file—the tool we used to identify likely deceased individuals. In addition, FEMA officials stated that the agency's validity check does not confirm that the date of birth is consistent with the name and SSN on the application, which we were able to do by using SSA's EVS. Such differences could explain why we were able to identify more potentially fraudulent transactions, such as recipients being potentially deceased or as having SSNs that did not match the associated name and date of birth as registered with SSA. As an assistance-providing federal agency, FEMA may have the authority to access SSA's more-comprehensive death file data.[38] FEMA may also be able to establish an information-sharing agreement with SSA to access EVS in order to verify the SSNs provided by IHP applicants.[39] However, in providing technical comments on a draft of this report in November 2014, FEMA stated that as of October 2014 it had determined that the agency may be able to crosscheck applicant data with SSA's full death file. FEMA further stated that it had not contacted SSA to explore the cost and feasibility of accessing the full death file or EVS, although officials are open to considering this option if it is found to be the best approach. Under IHP, FEMA is to ensure that recipients are eligible for assistance. Supplementing the verification check that the contractor uses with SSA data could enable FEMA to further identify potentially ineligible applicants and reduce the risk of improper or fraudulent payments.

FEMA's Internal Controls Helped Prevent Improper Payments to Most but Not All Applicants Using Duplicate Applicant Information

FEMA's National Emergency Management Information System (NEMIS) uses an automatic control to identify IHP applications containing the same identifying information as other applicants to preclude payment of duplicate benefits. FEMA officials explained that this control is designed to identify applicants who have submitted duplicative information in one or more of four fields on the application: SSN, address, phone number, and bank account number. For example, the control is designed to hold payments to an applicant using the same SSN as another applicant who received assistance until a FEMA official can manually review the application and determine whether payment is appropriate. FEMA officials document in NEMIS their rationale for approving payments in these cases.

Duplicative information in one or more of those four fields can suggest potential fraud. Specifically, it can indicate that an applicant is applying multiple times or that more than one member of a household is applying for IHP assistance, which is generally not allowed in the program.[40] For example, SSNs are assigned to individuals, so applicants using the same SSN are likely to be either one person with multiple applications or an instance of possible identity fraud. Duplicate bank accounts also indicate a higher risk that multiple applicants from one household are requesting assistance. For example, two individuals sharing a bank account are likely to share living expenses and income, as one might expect to see in a household. Applicants with duplicate identifying information in the address and phone number fields may pose a risk, for example, of more than one person applying for benefits for the same damage, such as to a vehicle or other personal property, for which only one person, alone or as head of household, is eligible.

We tested whether FEMA's control to identify applications containing the same identifying information as other applications functioned consistently and found that it identified most such applications. Specifically, we found that, of the 182,782 IHP recipients, 36,021 records shared duplicate SSNs, addresses, phone numbers, or bank accounts with one or more other applicants. FEMA's control flagged 35,818 (99 percent) of these 36,021 recipients as having duplicate information when the applications were first processed. According to FEMA, these payments were to be halted until FEMA staff manually reviewed the applications. FEMA provided several examples from NEMIS documenting FEMA staff evaluation of the duplicate information and their eligibility determination.

However, the control did not identify the remaining 203 duplicate applicants, who received financial assistance totaling about $1.7 million. We provided information on these missed duplicates to FEMA for review. FEMA officials provided the following explanations:

- Officials stated that the control should have identified 30 of these applicants. As a consequence, a FEMA official should have manually reviewed these applications before a payment was processed. FEMA officials are investigating these 30 cases and said that they will analyze why NEMIS did not flag the applications.
- In 125 cases, officials explained that the duplicate applications were filed in different states and thus assigned a different disaster number. FEMA assigns separate disaster numbers for each state affected by a single disaster as each state receives its own presidential declaration of disaster. Because FEMA's duplicate control does not compare applications across different disaster numbers in different states, FEMA's control did not identify these applications as having duplicate information. As a consequence, it faced increased risk of providing improper assistance.[41]

As of August 2014, FEMA had not provided explanations for the 48 remaining instances in which we identified applications with duplicate information that were not flagged by FEMA's system. FEMA officials were unable to explain why the control did not function properly for all of the cases we identified but said that they will examine the control further to determine why it functioned inconsistently. According to standards for federal information-system controls, evaluating controls can help reduce risk in government programs.[42]

Identifying the cause for the control's inconsistent performance to the extent that the benefits exceed the costs could better position FEMA to determine whether modifications to the control are appropriate to reduce the chance of it missing duplicates in future applications. Additionally, *Standards for Internal Controls in the Federal Government* states that risks should be analyzed for their possible effect in order to determine how to manage the risks.[43]

As such, an analysis of the assistance provided to recipients using duplicate identifying information who were not flagged by FEMA's internal controls, including the costs of options available to address the issue, may provide insight as to how to effectively manage this risk in the future.

In addition to testing whether FEMA's control consistently identified applications with duplicative identifying information, we independently analyzed the 36,021 recipients to identify what we considered "high-risk" recipients: These were recipients who FEMA's control flagged as a duplicate but for which FEMA subsequently provided assistance payments and that met our criteria for high risk. These criteria, as discussed in greater detail below, included any instance in which assistance was provided to (1) recipients using the same SSN, (2) recipients using the same bank account, and (3) recipients using the same address or phone number as well as other identifying information such as the same last name, which are indications that two individuals are of the same household.

Recipients with Duplicate SSNs

Because SSNs are assigned to individuals, applications with the same SSN are likely to be either one person with multiple applications or an instance of possible identity fraud. Our analysis found that 10 recipients received assistance through two registration accounts each, both of which used the same SSN and other identifying information. Collectively, the 10 recipients were recorded as 20 different individuals in NEMIS and obtained $180,725 in assistance.

We provided information about these cases to FEMA for explanation. In 5 cases, FEMA officials stated that the manual review had improperly allowed payment, and FEMA initiated efforts to further review and, as appropriate, recoup the assistance.

FEMA officials stated that a caseworker had not followed proper procedures in completing manual reviews for these 5 individuals. In four cases, FEMA officials permitted an individual to use a second registration account to obtain additional assistance for funeral expenses for relatives who passed away as a result of the storm. We discuss these cases in greater detail below.

In the last case, FEMA officials stated that the applicant withdrew the second application upon being informed of the duplication, and that, while a rental assistance check had been issued through the second registration, the individual would have been eligible for the additional payment had the individual continued to use the first registration, so recovery of payments was unnecessary.

Recipients with Duplicate Bank Accounts

Individuals sharing a bank account are likely to share living expenses and income, as one might expect in a household with cohabitants. However, only one member of a household is generally eligible to receive IHP assistance. We found 147 recipients out of the 36,021 duplicates who shared a bank account with at least one other recipient, receiving a total of about $1.2 million. Furthermore, of these 147 recipients, 81 had matching last names in addition to bank account numbers. In 22 of those cases, the address and phone number also matched, increasing the risk of an improper or potentially fraudulent payment (see fig. 3).

FEMA officials stated that 133 of the 147 recipients did not receive duplicative benefits. Payments to 4 recipients had already been identified as improper and recouped by FEMA, and an additional 10 were identified and submitted for recoupment review.

In addition, we identified nine bank accounts among these 147 recipients that received payments totaling more than $31,900, the maximum amount allowable per household. FEMA officials stated that payments were allowable for eight of these nine bank accounts. They further explained that they began seeking recoupment of payments to the other bank account in January 2013 when they discovered that the manual review had incorrectly determined to pay this applicant.

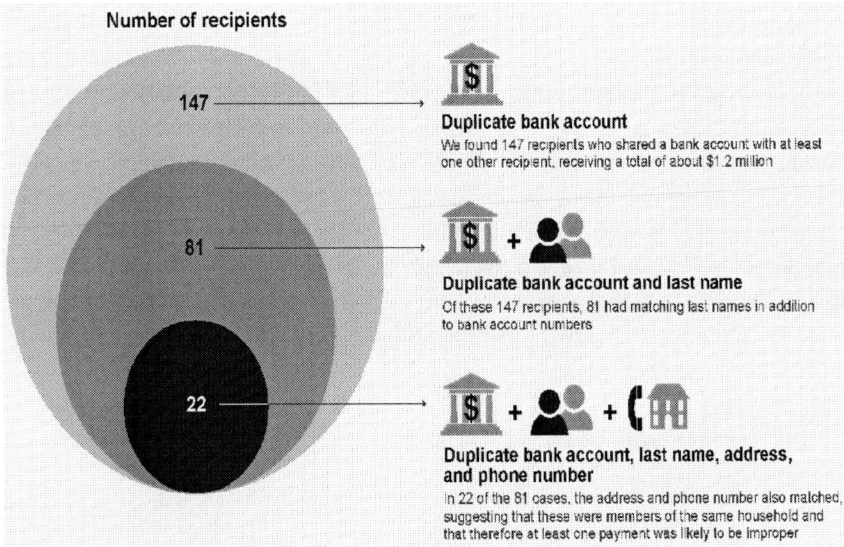

Source: GAO analysis of FEMA data.| GAO-15-15.

Note: As of October 2014, the Federal Emergency Management Agency (FEMA) reviewed payments to these 147 recipients, stating that 133 did not receive a duplication of assistance and that the remaining 14 recipients were either being reviewed for recoupment of payments or had already repaid the duplicative assistance. We could not independently confirm FEMA's conclusions without additional investigation for each payment.

Figure 3. Individuals and Households Program (IHP) Recipients Using Duplicate Bank Accounts.

Recipients with Duplicate Addresses or Phone Numbers and Other Risk Factors

In cases where we found recipients with duplicate addresses or phone numbers, we looked for additional risk factors that could indicate the applicants were part of the same household.[44] Specifically, we found 128 recipients using 66 addresses at which more than one person received home-repair assistance—a type of assistance that FEMA officials said should be provided to only one person per home regardless of how many people lived in a damaged home. This assistance totaled $1.3 million. FEMA reviewed the records of these 128 recipients and stated that 99 recipients did not receive duplications of assistance, while payments to 3 recipients had already been identified as improper and recouped by FEMA and an additional 26 were identified and submitted for recoupment review.

Additionally, we found 1,007 addresses at which multiple recipients cumulatively received more than the $31,900 maximum allowed per household. In 202 of those instances, totaling $7.6 million and including 418 recipients, applicants shared a last name with another recipient at that address, suggesting that the recipients may have composed a household (see fig. 4). FEMA's review of 200 of the 418 recipients concluded that 46 of these 200 recipients did not receive duplications of assistance and an additional 100 maintained renter and landlord relationships, so all parties were eligible for the assistance received. According to FEMA officials, the remaining 54 recipients were in recoupment review or had repaid the duplicative payments.

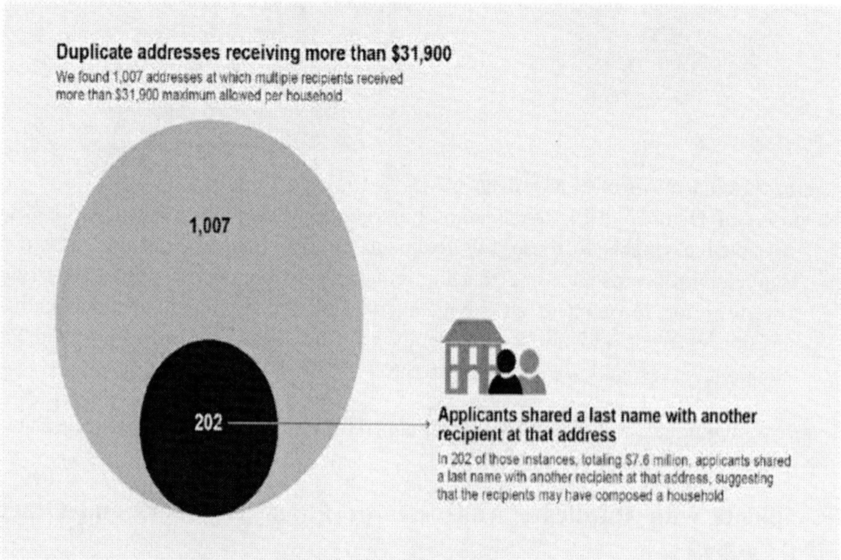

Source: GAO analysis of FEMA data.| GAO-15-15.

Note: $31,900 is the maximum allowable amount of assistance to an individual or household.

As of October 2014, FEMA reviewed 200 of the 418 recipients residing at the 202 addresses where recipients shared a last name and address, stating that 145 did not receive a duplication of assistance and that the remaining 55 recipients were either being reviewed for recoupment of payments or had already repaid the duplicative assistance. We could not independently confirm FEMA's conclusions without additional investigation for each payment.

Figure 4. Individuals and Households Program (IHP) Addresses at Which Multiple Recipients Received More than a Combined $31,900.

Receipts of Funeral Assistance through a Second Registration in the Same Household

As mentioned above, FEMA officials stated that there were 4 instances in which an individual applicant was allowed to use a second registration number to apply for and receive funeral expenses. However, we found 5 additional instances in which members of a single household were allowed to create a second registration account to receive additional assistance for funeral expenses.[45] The 9 cases involved 18 recipients who received a total of nearly $72,000 for funeral expenses. In these cases, individuals received as much as an additional $10,000 for funeral expenses, and in 3 cases the total amount of assistance exceeded the $31,900 limit per household. While this approach was not consistent with federal law, the decision to provide additional aid was made for humanitarian reasons. In April 2014, FEMA clarified its guidance to explicitly state that a second registration could not be used for funeral expenses, as that could allow a household to exceed the $31,900 assistance cap.

FEMA Could Enhance Oversight of Assistance for Flood-Related Damage

Individuals that receive IHP assistance for flood-related damage in a disaster and live in certain flood zones are generally required to purchase and maintain flood insurance as a condition of receiving assistance from the IHP for flood-related damage.[46] An individual was generally not eligible to receive IHP assistance for flood-insurable items that incurred flood-related damage from Hurricane Sandy if the damaged property had a flood-insurance requirement from a prior disaster but the owner did not comply with this requirement and did not maintain a policy.[47]

FEMA uses information from two NFIP-related databases to populate fields in its IHP database to flag IHP applicants who are potentially ineligible for assistance for flood-related damage.[48] Table 4 shows the definitions of the flag FEMA uses to document flood-insurance requirement compliance and, by extension, applicant eligibility to receive flood-related assistance.

Table 4. Definitions of Flood-Insurance Compliance Flag in Federal Emergency Management Agency (FEMA) Database

Flagged as "Yes" (compliant)	Flagged as "No" (noncompliant)
Damaged property has a flood-insurance requirement and the policy is maintained;[a] or Damaged property has no flood-insurance requirement.	Damaged property has a flood-insurance requirement that is not maintained.

Source: FEMA. | GAO-15-15

[a]An applicant is considered compliant if there is an active flood-insurance policy at the time the damage occurs for at least the amount of assistance received during a previous disaster.

According to our analysis of the 182,782 IHP recipients, we found that 39,766 received about $385 million in assistance for flood-related damage.[49] However, FEMA flagged 1,322 of these 39,766 individuals as being noncompliant with an existing flood-insurance requirement and provided these individuals with about $10.5 million in assistance for flood-related damage.

FEMA reviewed a nongeneralizable sample of 63 of these noncompliant cases to explain why these applicants received assistance for flood-related damage. FEMA officials determined that in 26 of the cases, the recipient was compliant at the time of the assistance payment. In another 31 cases, the recipient was noncompliant at the time of Hurricane Sandy, but received assistance for items not covered by flood-insurance, which constitute eligible payments. In the remaining 6 cases, the assistance for flood-related damage was paid in error and is eligible for recoupment. The officials further explained that many of those flagged as noncompliant were individuals whose property did not have a flood-insurance requirement prior to Hurricane Sandy, and the noncompliant flag appeared after receiving IHP assistance for flood-related damage. According to the officials, the fields that are populated from the NFIPrelated databases are updated on a continuous basis. FEMA officials explained that what likely happened with most of the 1,322 applicants was that their damaged property was flagged as compliant at the time the assistance payment for flood-related damage was made, but subsequent to the payment, a data change to the applicant's file may have prompted the system to recheck the NFIP-related databases—and because there was no indication of the purchase of their newly required floodinsurance—the applicant information was updated and appeared in the IHP database as noncompliant. We could not independently corroborate the officials' explanations because the design of the compliance flag in the IHP database did not allow us to verify the applicant's flood-insurance status at the time the payment was made. Internal control standards require controls and transactions to be clearly documented and documentation to be readily available for examination.[50] The design of the compliance flag prevents FEMA or other reviewers from determining, after the fact, whether a noncompliance flag meant that an applicant was ineligible for assistance for flood-related damage after Hurricane Sandy. FEMA officials told us that currently there is not a way to tell through review of an applicant's file alone what an applicant's compliance status was at the time of the IHP assistance payment. As a result, when we presented FEMA officials with the 1,322 records that indicated a noncompliant applicant received IHP assistance for flood-related damage, FEMA officials could only determine whether a particular applicant was in fact eligible for the assistance payment for flood-related damage by reviewing an external NFIP-related database. The need for this extra step to review these transactions makes it more difficult to determine how much IHP assistance might have been improperly paid. FEMA officials also stated that the interaction between the IHP and NFIP databases was based on legacy systems that are in need of updating, which they are exploring. Specifically, FEMA officials told us that they are currently developing a budget for acquiring new information-technology capabilities and, in the agency's November 2014 written response to this report (see app. IV), produced a set of development milestones to improve this compliance flag by May 31, 2016. FEMA officials stated that they were supportive of an enhancement to improve the current system's ability to track flood-insurance compliance and provide additional assurances that recipients were, in fact, eligible for IHP assistance for flood-related damage. Updating the legacy systems could be a good opportunity to redesign the compliance flag in the IHP system so that it will preserve the applicant's compliance status at the time of the IHP assistance payment.

FEMA and the States Faced Challenges in Obtaining and Sharing Information Needed to Prevent Payments from Potentially Overlapping Sources

We found that FEMA and state governments did not have the information necessary to determine whether FEMA was successful in preventing IHP recipients from receiving overlapping payments from FEMA's Sheltering and Temporary Essential Power (STEP) program. FEMA implemented a new intra-agency data-sharing agreement in April 2014 to help make future information sharing between these programs easier. In addition, FEMA does not have a mechanism to verify whether individuals have homeowners insurance, which would limit IHP payments for damages also covered by private insurance.

Limited STEP Data Posed a Challenge in FEMA's Efforts to Design Controls and Determine Whether STEP Recipients Also Received IHP Assistance

FEMA's guidance states that individuals who participate in the STEP pilot program are generally ineligible for IHP rental assistance once STEP-related work is completed, but FEMA did not have controls in place to limit IHP assistance from being offered to STEP recipients. FEMA's evaluation of the STEP program could help develop these controls by first collecting data from states and localities that implemented STEP. Specifically, FEMA asked that localities provide data on STEP participants by September 30, 2014, and FEMA plans to use the data to evaluate the pilot program, including lessons learned. However, according to the agency's analysis plan, this information will not permit FEMA to determine how many STEP recipients were living in TSA housing or receiving IHP rental assistance after STEP work was completed on their residence, and several challenges may hinder its efforts.[51]

When it initiated STEP in the weeks following Hurricane Sandy, FEMA did not provide guidance to states and localities that would help it obtain complete information about the participants in the STEP program. Specifically, FEMA's guidance for data collection directed the localities that implemented STEP to provide FEMA and the appropriate state with a list of residences that received STEP assistance, including the owner's/occupant's name, residential address, and IHP registration ID, if applicable.[52] However, FEMA did not direct the localities to collect additional data elements on residents served, such as dates of service, which would have allowed FEMA to determine whether individuals were receiving rental assistance or remained in TSA lodging after repairs on their home were complete.[53] As a result, localities defined the required information differently and collected different types of data outside of the required fields. For instance, FEMA officials told us that some localities collected names and addresses of participants but not the dates on which repairs were begun and finished. In addition, localities defined service dates differently. For example, some localities defined the start date as the date on which the repair work was initiated, while other localities defined it as the date of the initial assessment. FEMA has identified the absence of consistent data as a challenge for its analysis and its ability to understand whether STEP recipients continued to receive rental or other potentially duplicative assistance after repairs were complete.

FEMA officials also told us that one obstacle to collecting STEP data for comparison to TSA and IHP data is that, prior to Hurricane Sandy, they did not take the steps under the Privacy Act of 1974 to allow them to collect more personally identifiable information, such as SSNs or dates of birth, on individuals who received assistance through Public Assistance programs like STEP. As a consequence, FEMA could not identify and compare individuals who received STEP assistance with those who received TSA and IHP. FEMA officials also stated that their ability to request recipient addresses from local jurisdictions and compare them against data on individuals receiving other types of assistance in real time was inhibited because they did not have this authority immediately after Hurricane Sandy.

In response to these issues, in March 2014, FEMA published a new System of Records Notice, as required by the Privacy Act, on data sharing to make information sharing between programs easier.[54] Specifically, under the new rule, which took effect in April 2014, in future disasters FEMA will be able to instruct localities to collect and share relevant information on properties and individuals receiving assistance through Public Assistance efforts such as STEP, and FEMA will be able to compare this to assistance provided through Individual Assistance. FEMA officials expect the new data rule will help them to more accurately identify and monitor private properties that have been damaged by a disaster and are in need of Public Assistance for purposes of preventing duplication of benefits. According to the officials, FEMA has set up a committee to explore ways to maintain and share this data by developing a comprehensive information-technology system to house relevant disaster data, including information gathered by IHP and Public Assistance so that different entities within FEMA can access it. This committee should provide FEMA an opportunity to identify the specific fields it will need to both evaluate and prevent potentially duplicative sources of assistance from pilot programs, like STEP, when it implements such programs in the future.

FEMA Made Payments at Risk of Being Improper Because It Did Not Have a Mechanism to Verify Whether IHP Recipients Had Homeowners Insurance

IHP provides disaster survivors with home-repair funds for damages not covered by private homeowners insurance, as federal law prohibits FEMA from providing assistance for damage covered by private insurance. To determine whether applicants carry private insurance, FEMA relies on self-reported data from each applicant. Specifically, applicants must identify any insurance policies they have during the registration process.

In order to identify properties likely to have been covered by a homeowners insurance policy at the time of the storm, we examined IHP home-repair assistance provided to addresses that held mortgages backed by Fannie Mae, Freddie Mac, or FHA. The federally related mortgages of these three entities require the homeowner to maintain private homeowners insurance on the property, according to officials at those agencies. Our analysis of the 182,782 IHP recipients found that 40,770 were homeowners who received assistance for home repair or personal property damage (excluding certain flood-related damage) and of these, 3,718 reported that they did not have any private insurance.[55] Of the 3,718 recipients, we found that 534 (14.4 percent) received assistance for damage to a property that had a federally backed mortgage that required private homeowners insurance as of October 2012. FEMA provided over $2.3 million in assistance for home repair or personal property damage to these individuals (see table 5 below).

We consider these payments to be at risk because these homeowners were likely to have insurance that typically covers this type of damage, although they reported that they did not. However, it is not possible to determine whether these matches definitively identify improper payments without reviewing the facts and circumstances of each case. Under certain circumstances, FEMA can appropriately provide assistance for home repair or personal property to individuals with private home insurance policies. For example, FEMA may provide IHP assistance for damage that is not covered by a particular private insurance policy or for damage that exceeds the limit on a private insurance policy.[56] In addition, if an insurance payment is delayed by more than 30 days, FEMA may provide applicants with home-repair payments, which are considered an advance and must be repaid to FEMA once an insurance settlement is received. FEMA reviewed 55 of these cases and stated that 32 were likely appropriate because the assistance was for damage that likely would not have been covered by private insurance. Nevertheless, these payments are at higher risk of being improper if the recipient has an unreported homeowners insurance policy.

Table 5. Individuals and Households Program (IHP) Home-Repair and Personal-Property Assistance for Recipients That Reported No Private Insurance but Held Federally Backed Mortgages That Require Insurance Coverage

Mortgage holder	Number of recipients thatreported not havingprivate insurance andreceived IHP assistance	Assistance received[a] (dollars)
Federal National Mortgage Association (Fannie Mae)	81	$635,746
Federal Home Loan	43	166,205
Mortgage Corporation (Freddie Mac)		
Federal Housing Administration (FHA)	410	1,522,798
Total	534	$2,324,749

Source: GAO analysis of Federal Emergency Management Agency (FEMA), Fannie Mae, Freddie Mac, and FHA data. | GAO-15-15

Note: As of October 2014, FEMA reviewed 55 of the 534 recipients, stating that 32 did not receive a duplication of assistance. We could not independently confirm FEMA's conclusions without additional investigation for each payment.

[a]Excludes assistance to homeowners for flood damage to properties located in a special flood hazard area because such flood damage is typically not covered by private homeowners insurance.

GAO's fraud control framework states that information provided by applicants for government programs should be validated against government or other third-party sources.[57] There are a number of options that may help FEMA better evaluate whether IHP applicants have homeowners insurance policies. Implementation costs and the extent to which self-reported data could be validated vary among these options. For instance, to mitigate instances in which an applicant was not aware that he or she had homeowners insurance or did not understand the question, FEMA could ask whether applicants have a mortgage, since most mortgages require the mortgagee to maintain homeowners insurance. Although FEMA officials noted that having a mortgage did not automatically mean that individuals have homeowners insurance, it would allow the agency to flag applicants who said that they had a mortgage but no homeowners insurance for further verification of lack of insurance. Alternatively, FEMA has the legal authority to solicit this information from housing agencies, which means it may be able to periodically check applicant addresses against Fannie Mae's, Freddie Mac's, and FHA's federally related mortgages, as we did. This control would not prevent all potentially improper payments, as not all mortgages are federally related, but could provide an alternative to contacting individual insurance companies while allowing FEMA to recoup some funds that were provided for home repairs already covered by homeowners insurance. FEMA has not explored collaborating with federal entities that guarantee mortgages, such as FHA, Fannie Mae, or Freddie Mac. Lastly, public-private models for data sharing have proved successful for other agencies. For instance, MITRE, a nonprofit corporation that operates federally funded research and development centers chartered by Congress, recently assisted a federal agency in exchanging sensitive information with several private companies from a competitive industry. MITRE collected, held, and conducted analysis that all participants agreed to. It then provided the results to the parties while protecting both sensitive federal data and business competitive data that the parties would not want to share with one another.[58] A similar approach could help FEMA more systematically obtain data from private insurance companies. FEMA officials told us that they have the legal authority to solicit information from private insurance companies and have assessed possible mechanisms to check IHP and Public Assistance applicant information with data of third parties such as insurance companies in order to prevent a duplication of benefits. However, the agency did not implement this solution because the proposed mechanism raised privacy concerns and may have used incomplete information. Evaluating options, including the costs and feasibility, for verifying the accuracy of self-reported information FEMA received on whether applicants have private homeowners insurance could better position FEMA in making decisions about future assistance payments to help prevent duplication of payments to those applicants with homeowner insurance benefits.

CONCLUSION

In the 7 years between Hurricane Katrina and Hurricane Sandy, FEMA took steps to improve its ability to prevent ineligible applicants from receiving assistance following a disaster. For example, FEMA's controls for verifying identity, residence, and other key identifying information have helped identify high-risk applications and reduced the number of ineligible individuals from receiving payments, compared with previous disasters. However, additional opportunities exist to further enhance FEMA's oversight of its assistance. Given the limited amounts FEMA has been able to recover from improper payments for Hurricane Katrina (approximately 10 percent of the original amount has been recovered), strengthening its preventative, up-front controls is a better way to protect taxpayer resources than attempting to recover improper payments after they are made. For example, without collaborating with and obtaining additional data from SSA, particularly data from EVS and the full death file that SSA maintains, FEMA risks paying ineligible applicants because its automated identity-verification control would remain incomplete. Additionally, assessing the feasibility of options to enhance FEMA's automated control that flags applications with duplicate identifying information may help management select cost-effective ways to further reduce the risk of improper payments. With regard to flood assistance, modifying data fields in its IHP database would allow FEMA's management to better assess the extent to which only eligible applicants received assistance for flood-related damage. Unless FEMA leverages these opportunities to make changes, the potential for improper assistance payments like those we identified in this report will likely persist.

FEMA continues to pilot new ways of providing assistance to disaster survivors, such as the STEP program. FEMA encountered challenges obtaining data from states on the delivery of STEP assistance after Hurricane Sandy in part because it did not request necessary data fields and, at the time, it did not have the authority to compare certain types of data on individuals receiving support from its Individual Assistance and Public Assistance programs. By determining data requirements and developing a process for comparing these data across programs before future disasters, FEMA would be better positioned to both evaluate STEP and ensure in future disasters that individuals whose homes are made habitable do not continue to receive duplicative types of assistance, such as rental assistance.

Finally, FEMA provided home-repair assistance to some applicants who may have inaccurately indicated that they did not have homeowners insurance, meaning that FEMA would have paid for damage already covered by a private source. This risk stems from FEMA's reliance on self-reported data for a single question—whether the applicant has private insurance—that is part of the application process. FEMA has begun exploring a mechanism for validating these data. Other approaches—each with varying levels of cost-effectiveness—could provide greater assurance that FEMA has accurate and reliable information on whether homeowners truly need its assistance or could turn instead to their private insurance to recover damages.

RECOMMENDATIONS FOR EXECUTIVE ACTION

To help FEMA prevent improper payments, we recommend that the Administrator of FEMA take the following three actions:

- Collaborate with SSA to assess the cost and feasibility of checking recipient SSNs against EVS and the full death file to more accurately identify recipients who used SSNs that were ineligible or belonged to likely deceased individuals, document the results of this assessment, and if determined to be cost-beneficial take steps to implement a partnership to use SSA data.
- Assess the cost and feasibility of addressing limitations in FEMA's control for identifying duplicate information in applications in high-risk data fields—such as SSN, bank-account information, address, and phone number—that may currently allow individuals or households to improperly receive multiple payments, and if determined to be cost-beneficial take steps to address the system design limitation.
- As part of updates to legacy systems, redesign the compliance flag in the IHP system to clearly identify and document applicants' compliance with NFIP requirements at the time when assistance for flood-related damage was provided through IHP.

To facilitate more-effective data sharing with state-level partners and enhance FEMA's ability to prevent duplicative benefits, we recommend that the Administrator of FEMA take the following action:

- As part of its committee that is implementing enhanced data-sharing between Public Assistance and Individual Assistance programs, establish data-reporting requirements for states, including specific fields needed and a standard process for comparing information across programs, including IHP and STEP, to better position FEMA to evaluate such pilot programs and to help prevent potential duplicative payments.

Finally, to prevent assistance that duplicates homeowner insurance benefits, we recommend the Administrator of FEMA take the following action:

- Evaluate options, including costs and feasibility, to identify an approach for verifying the accuracy of self-reported information FEMA receives on whether applicants have private homeowners insurance. Such options could include posing additional questions to applicants, sharing data with federal agencies to identify federally backed mortgages, or developing a data-sharing approach with private insurance companies.

AGENCY COMMENTS AND OUR EVALUATION

We provided a draft of this report to the Secretary of Homeland Security for comment. In its written comments, summarized below, the Department of Homeland Security (DHS) concurred with all five recommendations. DHS also described ongoing or planned actions, and provided a timeline for addressing the recommendations. DHS further stated that FEMA had already implemented the fifth recommendation; however, we continue to believe that additional action is needed to fully address that recommendation.

In response to our first recommendation, to collaborate with SSA to assess the cost and feasibility of checking recipient SSNs against EVS and the full death file, DHS said that FEMA would engage with SSA to assess the cost and feasibility of using EVS and the full death file, as well as work with its third-party vendor to assess the feasibility of improving SSN validation results. The agency expects to complete this work by December 31, 2015.

In response to our second recommendation, to assess the cost and feasibility of addressing limitations in FEMA's control for identifying duplicate information in applications in high-risk data fields that may currently allow individuals or households to improperly receive multiple payments and take steps to address the system design limitation if cost-beneficial, DHS said that FEMA would work with developers to ascertain why certain duplicate applications were not flagged, anticipating a completion date by January 2015. DHS also stated that FEMA would consider appropriate system improvements based on the review and that a schedule of system changes would be developed according to the availability of funds.

In response to our third recommendation, to redesign the compliance flag in the IHP system to clearly identify and document applicants' compliance with NFIP requirements at the time when assistance for flood-related damage was provided through IHP, DHS said that FEMA's Recovery Directorate would identify a means for recording changes to applicants' compliance with NFIP requirements. According to DHS, the changes are to be easily viewed and queried to show applicants' status throughout the phases of the application process. FEMA estimated that work would be completed by May 31, 2016.

These actions, if implemented effectively, would address the intent of our recommendations.

In response to our fourth recommendation, to establish data-reporting requirements for states, DHS stated that FEMA would enhance its information-technology infrastructure to better facilitate bidirectional data sharing with states. FEMA started this effort in 2013 and plans to continue this work through September 30, 2016, in developing a datacenter to process this data exchange. While this action may help facilitate the sharing of data between FEMA and states, additional steps will be required to clarify at the outset of the response to a given disaster what data states are to collect and when the data will be shared with FEMA.

In response to our fifth recommendation to evaluate options to identify an approach for verifying the accuracy of self-reported information FEMA receives on whether applicants have private homeowners insurance, DHS stated that FEMA recognized the value in obtaining insurance data for IHP applicants and described the efforts FEMA has taken to-date to obtain this information. According to DHS, efforts to obtain the data from private-sector companies were not successful because FEMA is precluded from complying with the private-sector companies' expectations of receiving personally identifiable information about applicants in return from FEMA. Additionally, DHS stated that FEMA determined that access to insurance information from individual private insurance companies was not cost-effective, although DHS did not provide evidence to demonstrate that an analysis of the costs was completed. DHS stated that FEMA had considered acquiring data from other federal agencies to determine whether IHP recipients had federally backed mortgages, as we did in our analysis. According to DHS, FEMA contends that such analysis would only cover a small percentage of the insured population, yet we determined that nearly one-third of recipients of home repair or personal property assistance through the IHP after Hurricane Sandy had a federally backed mortgage. We believe that, given the inherent unreliability of self-reported data, adopting a mechanism to identify risk in a meaningful portion of applications warrants serious consideration. Further, though DHS states that an applicant's mortgage status does not influence eligibility for IHP assistance, we contend that mortgage status could serve as an appropriate flag for additional follow-up from a manual review of applications in which applicants claim not to have private home insurance. DHS also states that including one additional self-reported question on the IHP application regarding whether applicants have a mortgage would increase registration time and require approval from the Office of Management and Budget. One additional question should not significantly increase respondent burden. In addition, a question about mortgage status could provide an additional data point to assess the validity of responses to the current question. As a consequence, we reiterate our recommendation that FEMA evaluate these costs and potential benefits of adding a second question regarding whether an applicant has a mortgage as a flag for potentially inaccurate statements about insurance coverage. We appreciate the administrative burden associated with changes to the applicant form and suggest that FEMA could incorporate this change along with one of its future updates. While the percentage of applications in which this issue is a concern ultimately may not be large, the cumulative dollar amount of such potentially improper or potentially fraudulent payments from this disaster and future disasters could be significant. We consider this recommendation open until FEMA conducts additional analysis of the costs and benefits of these methods in screening applicants for misstatements about their private home insurance status.

FEMA also provided technical comments, which we incorporated as appropriate. In its technical comments, FEMA clarified several terms and provided additional details related to its policies and programs. FEMA also requested that the report reflect the agency's ability to provide supporting evidence pertaining to the at-risk payments we identified, which is reflected in the report. Additionally, FEMA suggested that some of the figures we report related to FEMA's supporting evidence did not match its calculations. We reviewed these comments but determined that the report correctly summarizes the results of our analysis and the evidence that FEMA provided.

Seto J. Bagdoyan
Director, Forensic Audits and Investigative Service

APPENDIX I: OBJECTIVES, SCOPE, AND METHODOLOGY

This report examines (1) the extent to which the Federal Emergency Management Agency (FEMA) implemented controls to help prevent Individuals and Households Program (IHP) payments that are at risk of being improper or potentially fraudulent and (2) what challenges, if any, FEMA and states faced in obtaining and sharing information to help prevent IHP payments from duplicating or overlapping with other sources of assistance in its response to Hurricane Sandy.

To assess the extent to which FEMA implemented controls to help prevent payments that are at risk of being improper or potentially fraudulent, we reviewed FEMA's controls designed to prevent individuals from fraudulently receiving disaster assistance, reviewed IHP data of applications and assistance provided in response to Hurricane Sandy, and interviewed FEMA officials. FEMA provided its IHP records as of November 2013, the most current and complete data available at the time of our request. The data contained records for all applicants requesting assistance for damages related to Hurricane Sandy, as well as the type and amount of IHP assistance payment received. To identify indications of payments to potentially ineligible recipients, we performed extensive testing of these data to determine whether there were (1) records with duplicate Social Security numbers (SSN), addresses, bank account numbers, and phone numbers and (2) payments to potentially ineligible recipients, such as those that were identified in other data systems as having invalid SSNs or as likely deceased or incarcerated at the time an assistance payment was received. To identify whether any of the award recipients had potentially inaccurate or invalid personal identification information (SSN, name, birth date), we submitted this information for verification to the Social Security Administration's (SSA) Enumeration Verification System (EVS). The EVS system provides information on invalid (never issued) SSNs and instances where there are mismatches between SSN, name, and birth date. Where there are mismatches between SSN and name or SSN and birth date, EVS may provide one or more suggested SSNs based on the applicants' name and birth date. Where EVS provided these alternatives, we compared the suggested SSN to the SSN in the IHP data to determine the likelihood that a mistake had been made in transcribing them. In some cases, we found an EVS suggested SSN to be very similar to the SSN on the corresponding application, such as a match on 8 of 9 digits, leading us to conclude that the SSN was a match. However, if the mismatch was greater than 2 digits, we considered the SSN unverifiable with SSA's records. We defined invalid SSNs as those that EVS determined had never been issued.

To identify IHP recipients who may have used the identity of a likely deceased individual to receive IHP assistance, we matched FEMA IHP recipient data to the SSA's full death file by SSN to identify cases where the IHP application date was after the reported date of death. The full death file contains all of SSA's death records, including state-reported death information. We used a copy of SSA's full death file that was current as of May 2014, the most recent version available at the time we performed the analysis. We identified matches in two ways. In the first, we compared the SSNs that IHP recipients used in their application with the SSNs of those listed on the full death file. In the second, we examined the results of our analysis using the EVS system, which verified that IHP applicant information matched SSA's records. EVS identified some IHP recipients whose SSNs did not correspond to their name or date of birth. For some of these IHP recipients with mismatched identifying information, EVS provided a valid SSN for the name and date of birth of the applicant. For some of the flagged SSNs, EVS provided one or more valid SSNs that matched on the name and date of birth (or just on the name). We checked these EVS-recommended SSNs against the full death file. We used the EVS-recommended valid SSN to compare these IHP recipients with the SSNs of those listed on the full death file. For cases where EVS recommended more than one SSN for an IHP recipient, we only counted it as a full death file "match" if all the EVS-recommended SSNs matched with an individual on the death file. To identify award recipients with SSN's associated with incarcerated individuals at the time of FEMA assistance, we matched FEMA IHP data to state department of corrections prisoner databases from New York, New Jersey, and Connecticut.[1] Prisoner data included individuals incarcerated on the day that Hurricane Sandy made landfall, October 29, 2012, and at any time in the following 6 months (through April 29, 2013). We identified records for which the FEMA assistance recipients' SSN, name, and birth date, or name and address matched that of a person who was incarcerated from October 29, 2012 through April 29, 2013. We conducted an analysis of the matches to eliminate those individuals who could have qualified for rental assistance prior to or following their incarceration. We also eliminated cases that appeared to be a result of two individuals sharing a name or address but having different dates of birth and SSNs. It is not possible to determine from data matching alone whether these matches definitively identify recipients who were deceased or incarcerated without reviewing the facts and circumstances of each case. For example, it is possible that individuals can be erroneously listed in the full death file. Similarly, a recipient may have an SSN, name, and date of birth similar to an individual in state prison records. Alternatively, our matches may also understate the number of deceased or incarcerated individuals receiving assistance because matching would not detect applicants whose identifying information in the FEMA data differed slightly from their identifying information in other databases. To assess FEMA's oversight of IHP assistance for flood-related damage, we analyzed FEMA's IHP data to determine the number of recipients and amount of assistance received for flood-related damage from Hurricane Sandy whose property carried a requirement to purchase and maintain flood-insurance on that damaged property for future disasters. We examined how FEMA determined and tracked the eligibility of IHP recipients to receive assistance for flood-related damage based on their compliance with an existing flood-insurance requirement from a previous disaster. We analyzed the IHP data to identify how much flood-related assistance, if any, FEMA provided to IHP recipients it had flagged as noncompliant with an existing flood-insurance requirement. An applicant who was noncompliant with a flood-insurance requirement could still be eligible for IHP flood-related assistance to repair items that are not covered by flood insurance, such as wells, sewers, or septic systems, as well as for types of assistance that are not dependent on whether an applicant has flood insurance, such as rental assistance or assistance with medical, dental, childcare, or disaster-related funeral costs.

We assessed the accuracy and reliability of FEMA's IHP data by interviewing agency officials knowledgeable about the data system and by obtaining from the agency written responses regarding (1) the agency's methods of data collection and quality-control reviews, (2) practices and controls over data-entry accuracy, and (3) any limitations of the data. In addition, we conducted electronic testing of the data, including checks for missing, out of range, or logically inaccurate data. We determined that the data were sufficiently reliable for the purposes of our report.

We also assessed the reliability of SSA's full death file and prisoner databases provided by departments of corrections in Connecticut, New Jersey, and New York by reviewing relevant documentation, interviewing knowledgeable agency officials, and examining the data for obvious errors and inconsistencies. We concluded that each database was sufficiently reliable for the purposes of this report. In addition to data analysis, we identified criteria for FEMA's actions by examining policies, laws, and guidance, including the policy memos and applicant guidance published by FEMA for the National

Flood Insurance Program (NFIP) and IHP. We reviewed changes to FEMA's controls since Hurricane Katrina and interviewed officials from FEMA, state governments of New York and New Jersey, and the city of New York involved in the recovery effort about the effectiveness of the controls and any likely opportunities for fraud. We also reviewed relevant legal materials, including the Post-Katrina Emergency Management Reform Act of 2006 (PKEMRA).[2] In addition, we referred to best practices cited by auditing and antifraud organizations, GAO's fraud control framework, and previous GAO work on information-systems controls.[3]

To examine the challenges, if any, FEMA and the states faced in obtaining and sharing information necessary to prevent duplicative payments from overlapping sources, we reviewed guidance about FEMA's Sheltering and Temporary Essential Power (STEP) pilot program to assess FEMA's ability to determine whether an applicant's home had been repaired through the STEP program prior to the cessation of sheltering assistance through the IHP. Finally, we interviewed federal, state, and local officials involved in the response to Hurricane Sandy about challenges faced in data-sharing efforts.

To identify individuals who may have received IHP assistance for damage from Hurricane Sandy that could be covered by private insurance companies, we requested that the Federal National Mortgage Association (Fannie Mae), and the Federal Home Loan Mortgage Corporation (Freddie Mac) compare the addresses of individuals listed who received home-repair or personal property assistance from IHP with their loan records for October 2012, the month that Hurricane Sandy made landfall, to identify addresses that had a mortgage backed by one of these two entities. We also obtained data on mortgages backed by the Federal Housing Administration (FHA) as of September 30, 2012 and compared these data with IHP data on individuals who received home-repair or personal-property assistance to identify addresses that had an FHA-backed mortgage. According to officials from these housing and mortgage organizations, mortgages backed by Fannie Mae, Freddie Mac, and FHA require that loan servicers (banks) ensure that mortgage holders maintain hazard insurance on the mortgaged property, so it is more likely that individuals who had a mortgage would also carry a hazard insurance policy. We determined whether any of the IHP recipients listed as homeowners who reported not having any private insurance received assistance for a damaged address that had a federally backed loan, and we analyzed the types and amounts of assistance received. We assessed the reliability of data from FHA, Fannie Mae, and Freddie Mac by interviewing agency officials knowledgeable about the data system and mortgage loans. In addition, we conducted electronic testing of the data, including checks for missing, out of range, or logically inaccurate data. We determined that the data were sufficiently reliable for the purposes of our engagement. However, it is not possible to determine whether these matches definitively identify improper payments without reviewing the facts and circumstances of each case.

We identified criteria for data sharing within FEMA and with external organizations by examining policies, laws, and guidance, including the policy memos and applicant guidance published by FEMA for the STEP program and IHP. We interviewed officials from FEMA, state governments of New York and New Jersey, and the city of New York involved in the recovery effort about the effectiveness of data-sharing efforts and controls to prevent duplication of benefits. We also reviewed relevant legal materials, including the Robert T. Stafford Disaster Relief and Emergency Assistance Act, as amended.[4] In addition, we referred to best practices cited by auditing and antifraud organizations and previous GAO work on program evaluation and GAO's fraud-control framework.[5] Finally, we assessed the reliability of data from Fannie Mae and Freddie Mac by interviewing agency officials knowledgeable about the data system and mortgage loans. In addition, we conducted electronic testing of the data, including checks for missing, out of range, or logically inaccurate data. We determined that the data were sufficiently reliable for the purposes of our engagement. We performed this audit from July 2013 through December 2014 in accordance with generally accepted government auditing standards. Those standards require that we plan and perform the audit to obtain sufficient, appropriate evidence to provide a reasonable basis for our findings and conclusions based on our audit objectives. We believe that the evidence obtained provides a reasonable basis for our findings and conclusions based on our audit objectives.

APPENDIX II: RECIPIENTS AND DOLLARS TO COUNTIES

The following maps represent a static version of the interactive graphic in figure 2, detailing Individuals and Households Program (IHP) assistance received by each county that was declared by the President to be a disaster area eligible for Individual Assistance programs. The President declares disaster areas at the request of the governor of each state. Although the Federal Emergency Management Agency submits a recommendation to the President, disaster declarations are ultimately at the President's discretion. Amounts may not total due to rounding.

Distribution of Individuals and Households Program Funds in Response to Hurricane Sandy for New Jersey

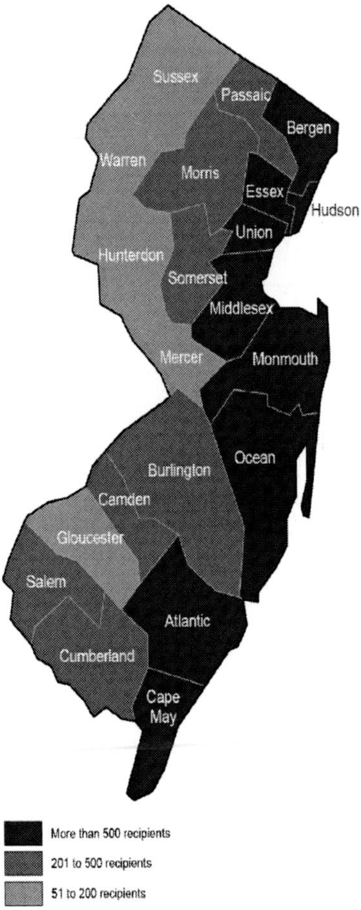

County	Number of recipients	Dollars to county
Atlantic	8,712	$44,657,370
Bergen	2,779	19,850,637
Burlington	245	1,110,082
Camden	421	1,193,538
Cape May	1,907	8,718,039
Cumberland	223	1,430,995
Essex	1,651	4,935,141
Gloucester	100	588,206
Hudson	5,172	27,667,126
Hunterdon	78	231,011
Mercer	198	499,172
Middlesex	2,980	19,422,435
Monmouth	11,537	85,579,069
Morris	264	594,614
Ocean	22,444	189,528,659
Passaic	270	688,018
Salem	207	1,218,273
Somerset	327	780,201
Sussex	151	374,941
Union	1,655	6,484,914
Warren	77	270,431
Total	**61,398**	**$415,822,873**

Source: GAO analysis of Federal Emergency Management Agency data. (data); | Map Resources (map). | GAO-15-15.

Distribution of Individuals and Households Program Funds in Response to Hurricane Sandy for Connecticut

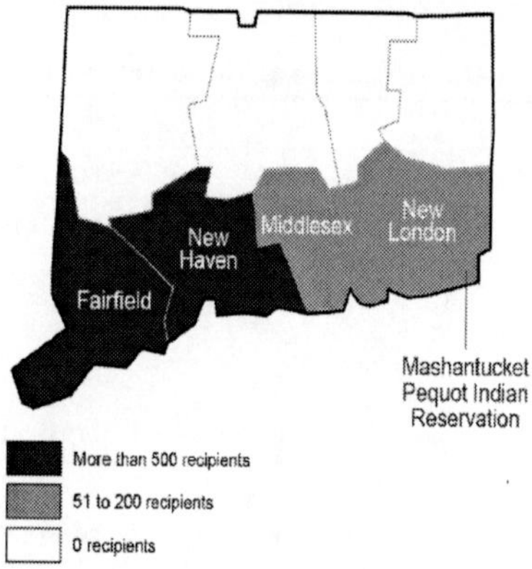

County	Number of recipients	Dollars to county
Fairfield	1,856	$9,726,137
Mashantucket Pequot Indian Reservation	1	248
Middlesex	124	553,567
New Haven	828	4,191,425
New London	160	613,317
Total	**2,969**	**$15,084,694**

Source: GAO analysis of Federal Emergency Management Agency data. (data); | Map Resources (map). | GAO-15-15.

Distribution of Individuals and Households Program Funds in Response to Hurricane Sandy for Maryland

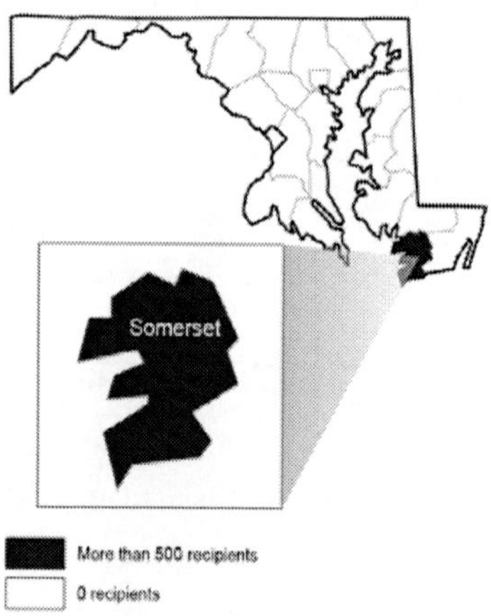

County	Number of recipients	Dollars to county
Somerset	732	$2,571,561
Total	**732**	**$2,571,561**

Source: GAO analysis of Federal Emergency Management Agency data. (data); | Map Resources (map). | GAO-15-15.

Note: Somerset County in Maryland was approved for Individual Assistance after an appeal, as it was the area with the greatest unmet need. The City of Chrisfield in Somerset County sustained the heaviest concentration of damage to its residences.

Distribution of Individuals and Households Program Funds in Response to Hurricane Sandy for New York

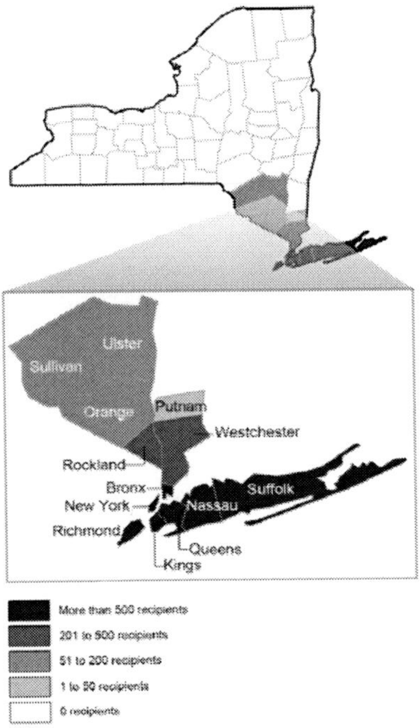

County	Number of recipients	Dollars to county
Bronx	653	$3,094,134
Kings	24,790	220,154,772
Nassau	39,696	325,512,835
New York	3,964	15,371,605
Orange	142	469,633
Putnam	48	183,337
Queens	27,086	250,781,103
Richmond	10,065	104,174,988
Rockland	473	3,242,520
Suffolk	10,066	80,031,647
Sullivan	76	283,604
Ulster	86	516,608
Westchester	433	1,695,853
Total	**117,578**	**$1,005,512,637**

Source: GAO analysis of Federal Emergency Management Agency data. (data); | Map Resources (map). | GAO-15-15.

Distribution of Individuals and Households Program Funds in Response to Hurricane Sandy for Rhode Island

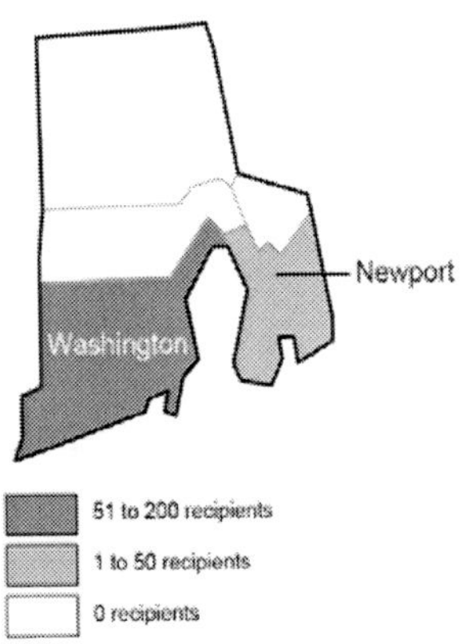

County	Number of recipients	Dollars to county
Newport	12	$69,650
Washington	93	351,691
Totals	**105**	**$421,341**

Source: GAO analysis of Federal Emergency Management Agency data. (data); | Map Resources (map). | GAO-15-15.

APPENDIX III: INDIVIDUAL AND HOUSEHOLDS PROGRAM (IHP) REGISTRATIONS REVIEWED BY THE FEDERAL EMERGENCY MANAGEMENT AGENCY (FEMA)

FEMA officials stated that at least $6.1 million of the payments we identified as being at-risk were deemed appropriate upon the agency's review of our matching results and related applicant files. Table 6 describes the results of FEMA's review for each type of risk for which payments were made, including the number of payments and dollar amounts of payments deemed appropriate as well as those recouped or under review.

Table 6. Summary of the Federal Emergency Management Agency's (FEMA) Review of the Individuals and Households Program (IHP) Assistance Identified as Being at Risk for Potential Fraud or Improper Payments

Type of risk for payments made	Number of IHP registrations identified as being at risk	Number of at-risk IHP registrations reviewed byFEMA	Number of IHP registrations FEMA deemed proper	Amount of at-risk payments FEMA deemed proper (dollars)	Number of at-risk IHP registrations with payments recouped or inreview	Amount of at-risk payments recouped or inreview (dollars)
Receipt of rental assistance while incarcerated	2	2	2	$7,328	0	$0
Social Security number (SSN) was invalid, did not match Social Security Administration's (SSA) records for other identifying information, or SSN belonged to likely deceased individual	2,610	77	67	781,028	10	120,472

Type of risk for payments made	Number of IHP registrations identified as being at risk	Number of at-risk IHP registrations reviewed byFEMA	Number of IHP registrations FEMA deemed proper	Amount of at-risk payments FEMA deemed proper (dollars)	Number of at-risk IHP registrations with payments recouped or inreview	Amount of at-risk payments recouped or inreview (dollars)
Duplicates not flagged by FEMA	203	203	n/a	n/a	n/a	n/a
SSNs used by multiple recipients	20	20	11	146,887	5	18,177
Bank accounts sed by multiple recipients	147	147	133	1,160,797	14	86,863
Addresses where more than 1 applicant received home-repair assistance	128	128	99	995,276	29	328,914
Address and last name used by multiple recipients and cumulatively receiving more than $31,900	418	200	146	2,699,422	54	888,856
Receipt of funeral assistance through a second registration in the same household	18	18	n/a	n/a	n/a	n/a
Receipt of home-repair assistance and claiming to lack private homeowners insurance while maintaining federally backed mortgage	534	55	32	259,418	23	87,381

Table 6. (Continued)

Type of risk for payments made	Number of IHP registrations identified as being at risk	Number of at-risk IHP registrations reviewed byFEMA	Number of IHP registrations FEMA deemed proper	Amount of at-risk payments FEMA deemed proper (dollars)	Number of at-risk IHP registrations with payments recouped or inreview	Amount of at-risk payments recouped or inreview (dollars)
Inaccurate flags for National Flood Insurance Program (NFIP) compliance	1,322	64	56	269,041	8	137,788

Source: GAO analysis of FEMA data. | GAO-15-15

n/a = not available

Note: Because some of the payments we identified as at risk appear in more than one category, the sum of the dollar amounts for each category in the table is greater than the sum of the entire population of at-risk payments, which counts the payments included in multiple categories only once.

We could not independently confirm FEMA's conclusions without additional investigation for each payment.

FEMA reviewed the 203 duplicates that were not flagged, but did not provide us with its assessment of the results for each applicant. Rather, FEMA presented its results in aggregate. As such, we do not have data about the dollar totals involved with cases FEMA deemed proper or attempted to recoup. Further, its explanations as to why the applicants were not flagged were not sufficient to conclude that the payments were proper. For example, FEMA claimed that 125 of the 203 duplicates resided in different states and, as such, FEMA did not consider the payments to be duplicative. We continue to believe that cross-state checks are necessary to detect potentially improper or fraudulent payments, as an individual with homes in two states should only be eligible for assistance related to the primary residence. Further, FEMA explained that 44 of the 203 duplicates had been flagged for manual review for cross-disaster checks (other disasters prior to Hurricane Sandy), but that would not preclude the records from being flagged as duplicates within this disaster, as we found.

Similarly, FEMA found that 16 of the 203 cases matched applications from disasters for which assistance was no longer available and were not flagged as cross-disaster duplicates because the matching disasters were closed. Again, our analysis found the duplicates within the data for only this disaster. FEMA confirmed that 30 applicants should have been flagged as duplicates and is investigating why the records were not flagged.

FEMA's review of funeral assistance using a second registration in the same household looked only for duplication of benefits instead of the extent to which second registrations were provided inappropriately within a household. While FEMA found no duplication of benefits for these payments, FEMA officials did not address our primary concern with these payments in its review and we did not include the results in this table.

End Notes

[1] For example, see GAO, Hurricane Sandy Relief: Improved Guidance on Designing Internal Control Plans Could Enhance Oversight of Disaster Funding, GAO-14-58 (Washington, D.C.: Nov. 26, 2013); and Catastrophic Disasters: Enhanced Leadership, Capabilities, and Accountability Controls Will Improve the Effectiveness of the Nation's Preparedness, Response, and Recovery System, GAO-06-618 (Washington, D.C.: Sept. 6, 2006).

[2] GAO, Hurricanes Katrina and Rita Disaster Relief: Improper and Potentially Fraudulent Individual Assistance Payments Estimated to Be Between $600 Million and $1.4 Billion, GAO-06-844T (Washington, D.C.: June 14, 2006); Expedited Assistance for Victims of Hurricanes Katrina and Rita: FEMA's Control Weaknesses Exposed the Government to Significant Fraud and Abuse, GAO-06-655 (Washington, D.C.: June 16, 2006); Hurricanes Katrina and Rita: Unprecedented Challenges Exposed the Individuals and Households Program to Fraud and Abuse; Actions Needed to Reduce Such Problems in Future, GAO-06-1013 (Washington, D.C.: September 27, 2006); and Hurricanes Katrina and Rita Disaster Relief: Continued Findings of Fraud, Waste, and Abuse, GAO-07-252T (Washington, D.C.: Dec. 6, 2006).

[3] In 2011, Congress passed the Disaster Assistance Recoupment Fairness Act of 2011, which allowed FEMA to waive recoupment of part of this total. The act permitted FEMA to waive debts under certain circumstances, but not if the debt involved fraud, the presentation of a false claim, or misrepresentation. FEMA stated that approximately $90 million was waived pursuant to the act.

[4] GAO-06-655, GAO-06-1013.

[5] GAO, Hurricanes Gustav and Ike Disaster Assistance: FEMA Strengthened its Fraud Prevention Controls, but Customer Service Needs Improvement, GAO-09-671 (Washington, D.C.: June 19, 2009). Our assessment was limited to undercover attempts to receive benefits and interviews with recipients of assistance; it did not include an examination of all assistance payments disbursed to survivors.

[6] Pub. L. No. 113-2, div. A, 127 Stat. 4 (2013). Amounts from the Disaster Relief Appropriations Act of 2013 represent the appropriations as originally enacted and not adjusted to account for sequestration.

[7] Since the data for IHP assistance was provided by FEMA in November 2013, IHP assistance has increased to 182,911 recipients as of August 2014.

[8] An improper payment is defined by statute as any payment that should not have been made or that was made in an incorrect amount (including overpayments and underpayments) under statutory, contractual, administrative, or other legally applicable requirements. It includes any payment to an ineligible recipient, any payment for an ineligible good or service, any duplicate payment, any payment for a good or service not received (except for such payments where authorized by law), and any payment that does not account for credit for applicable discounts. Improper Payments Information Act of 2002, as amended (codified at 31 U.S.C. § 3321 note). Office of Management and Budget guidance also instructs agencies to report as improper payments any payments for which insufficient or no documentation was found. Fraud involves an intentional act or representation to deceive with the knowledge that the action or representation could result in gain. An improper payment may be the result of fraud.

[9] The President declared that sections of 11 states (New Jersey, New York, Connecticut, Rhode Island, Maryland, Massachusetts, New Hampshire, West Virginia, Delaware, Pennsylvania, and Virginia) and the District of Columbia were adversely affected by Hurricane Sandy.

[10] SSA's full death file contains all of SSA's death records, including state-reported death information.

[11] We selected prisons in New York, New Jersey, and Connecticut because these three states were most affected by Hurricane Sandy.

[12] STEP was a pilot program intended to reduce the demand for other FEMA-provided housing, such as congregate shelters or TSA, and allow individuals to return to or remain in their homes. The program provided basic repairs to disaster-affected residences to make them habitable.

[13] PKEMRA was enacted as Title VI of the Department of Homeland Security Appropriations Act, 2007, Pub. L. No. 109-295, § 601, 120 Stat. 1355, 1394 (2006).

[14] Pub. L. No. 93-288, 88 Stat. 143 (1974) (codified as amended at 42 U.S.C. §§ 5121– 5207).

[15] GAO, Individual Disaster Assistance Programs: Framework for Fraud Prevention, Detection, and Prosecution, GAO-06-954T (Washington, D.C.: July 12, 2006); Federal Information System Controls Audit Manual (FISCAM), GAO-09-232G (Washington, D.C.: February 2009); Standards for Internal Control in the Federal Government, GAO/AIMD-00-21.3.1 (Washington, D.C.: November 1999); and Standards for Internal Control in the Federal Government, GAO-14-704G (Washington, D.C.: September 2014).

[16] FEMA regulations define Household to mean all persons who lived in the predisaster residence who request assistance, as well as any persons, such as infants, spouses, or part-time residents who were not present at the time of the disaster but who are expected to return during the assistance period.

[17] FEMA is required to adjust the maximum amount for assistance provided under the IHP annually based on the Consumer Price Index for All Urban Consumers.

[18] The end of the period of assistance for Hurricane Sandy was April 30, 2014, for New York, New Jersey, and Connecticut; May 3, 2014, for Rhode Island; and May 20, 2014, for Maryland.

[19] FEMA defines a congregate shelter as any private or public facility that provides refuge to evacuees but that normally serves a nonrefuge function, such as schools, churches, or stadiums.

[20] STEP was still a pilot program as of September 2014.

[21] For an overview of GAO's work on NFIP, see GAO, Economic Development: Overview of GAO's Past Work on the National Flood Insurance Program, GAO-14-297R (Washington, D.C.: Apr. 9, 2014).

[22] This estimate was based on a statistically significant sample and with a 95 percent confidence interval. GAO-06-844T.

[23] GAO-06-954T.

[24] GAO-06-1013.

[25] GAO, Hurricanes Katrina and Rita Disaster Relief: Continued Findings of Fraud, Waste, and Abuse, GAO-07-300 (Washington, D.C.: Mar. 15, 2007).

[26] GAO-09-671.

[27] In accordance with best practices from auditing and antifraud organizations, having a strong training program is an element of an effective fraud-prevention strategy. See, for example: Institute of Internal Auditors, American Institute of Certified Public Accountants, and Association of Certified Fraud Examiners, Managing the Business Risk of Fraud: A Practical Guide and Institute of Internal Auditors, Internal Auditing and Fraud. We did not evaluate this fraud prevention course or the fraud and abuse avoidance training as a part of this report.

[28] Department of Homeland Security, Office of the Inspector General, Improvements to Internal Controls for FEMA's Individuals and Households Program Registration Process, OIG-09-110 (Washington, D.C.: September 2009).

[29] This percentage was calculated using the actual results, not the rounded numbers displayed in the text.

[30] Office of Management and Budget guidance for fiscal year 2013 defines gross annual improper payments in a program exceeding (1) 2.5 percent of program outlays and $10 million of all payments or (2) $100 million regardless of the percentage of total payments as "significant improper payments."

[31] There are some instances in which an individual may be eligible to receive rental assistance despite being incarcerated when registering, such as when an individual is the head of a household applying on behalf of family members affected by the disaster.

[32] GAO-06-844T.

[33] We worked with state departments of corrections in New York, New Jersey, and Connecticut to access data on individuals who were incarcerated from October 29, 2012, through April 29, 2013.

[34] SSA has historically collected death information about SSN-holders so it does not pay Social Security benefits to deceased individuals and to establish benefits for survivors. SSA receives death reports from a variety of sources, including states, family members, funeral directors, post offices, financial institutions, and other federal agencies. We refer to SSA's complete file of death records as "the full death file." A subset of the full death file that may not include death data received by the states, which SSA calls "the Death Master File," is available to the public. For more information on SSA's death files, see GAO, Social Security Death Data: Additional Action Needed to Address Data Errors and Federal Agency Access, GAO-14-46 (Washington, D.C.: Nov. 27, 2013).

[35] GAO-06-844T.

[36] We identified 29 of the 45 recipients by matching the SSN used in their application with those listed on the full death file. The other 16 recipients were identified by using results of EVS, which identified some applicants as using an SSN that did not correspond to the applicant's name or date of birth. For some of the applicants with mismatched identifying information, EVS provided a valid SSN for the name and date of birth of the applicant. We used the EVS recommended valid SSN to compare these applicants with the full death file and found that the 16 recipients matched an individual who appeared on the full death file.

[37] Joint field offices are locations where federal, state, and local officials convene to administer their respective disaster-assistance programs.

[38] The Social Security Act requires SSA to share its full death file, to the extent feasible, with federal benefit-paying agencies for the purpose of preventing improper payments, if the agency reimburses SSA for its reasonable costs and the arrangement does not conflict with SSA's duties with respect to state data, as authorized by 42 U.S.C. § 405(r)(3). According to SSA officials, for FEMA to access these data, the agency must first make a formal request to SSA, explaining how it would use these data, and have agreements in place with SSA that outline the circumstances of a data-sharing agreement. SSA would then determine whether FEMA's proposed use of the information is in accordance with the allowable use outlined in the act. See GAO-14-46 for more information on this process.

[39] According to SSA officials, FEMA could initiate the process of establishing a reimbursable data-exchange agreement for SSN verifications by submitting a data-exchange request form found on SSA's website. SSA's Office of Data Exchange would then work with FEMA staff to identify how best to meet FEMA's needs. For example, in April 2007 the State

Department established an exchange agreement with SSA that allowed the State Department's electronic passport-processing system to automatically query EVS to verify that a passport applicant's SSN, name and date of birth match the records at SSA. See GAO, State Department: Pervasive Passport Fraud Not Identified, but Cases of Potentially Fraudulent and High-Risk Issuances Are under Review, GAO-14-222, (Washington, D.C.: May 1, 2014).

[40] According to FEMA regulations, a household may only register one time to receive IHP assistance, with some possible exceptions related to temporary housing assistance.

[41] In 2007, we recommended that FEMA conduct checks like this across disasters to ensure that damage covered for one disaster—for example, Hurricane Katrina—was not covered for a later disaster—for example, Hurricane Rita, which happened 1 month later. GAO-07-300.

[42] GAO, Federal Information System Controls Audit Manual (FISCAM), GAO-09-232G (Washington, D.C.: February 2009). We did not perform a complete review of information-system controls in this analysis.

[43] GAO/AIMD-00-21.3.1.

[44] There are legitimate situations in which a duplicate address or phone number would not suggest abuse. For example, two unrelated people, such as a homeowner and someone renting a room, may live at the same address and share a phone number and both may have been eligible for rental assistance in the event that the home was damaged in the storm.

[45] In two instances, recipients shared last names and bank accounts. In two other instances, recipients shared a last name, address, and phone number. And in the other instance, the recipients shared a last name and phone number.

[46] 42 U.S.C. §§ 4012a and 5154a; 44 C.F.R. §§ 206.110(k) and 206.113(b)(7) and (8). The flood-insurance requirement applies to IHP recipients residing in participating communities of the National Flood Insurance Program (NFIP) whose damaged property is located in a special flood hazard area. IHP recipients must purchase flood insurance, for at least the assistance amount, for the structure (if the IHP recipient is the owner) or for its contents (if the IHP recipient is a renter).The requirement extends to the flood-damaged property for the life of the property. If the property is sold, the flood-insurance requirement is passed on to the new homeowner. If a renter does not return to the property, he or she is no longer required to purchase flood insurance.

[47] An applicant who was noncompliant with a flood-insurance requirement could still be eligible for IHP assistance for flood-related damage to repair items that are not covered by flood insurance, such as wells or septic systems, as well as for types of assistance that are not dependent on whether an applicant has flood insurance, such as rental assistance or assistance with medical, dental, childcare, or disaster-related funeral costs.

[48] One NFIP-related database tracks information on flood-insurance requirements that result from receiving IHP flood damage assistance in a federal disaster; the other has information on all flood-insurance policies reported to the NFIP. The IHP database accesses information from these two NFIP-related databases based on the address of the damaged property. The NFIP-related databases are jointly managed by FEMA's Federal Insurance and Mitigation Administration and the Small Business Administration.

[49] As a consequence of receiving assistance for flood-related damage caused by Hurricane Sandy, these individuals will be required to purchase and maintain flood insurance for future disasters. This flood-insurance coverage must be at least the amount of IHP flood-related assistance received for damage from Hurricane Sandy.

[50] GAO/AIMD-00-21.3.1.

[51] GAO's guide for designing evaluations states that an evaluation gives an agency the opportunity to improve program performance and provides a useful tool to determine whether the program has resulted in the desired benefits for participants and the public. GAO, Designing Evaluations: 2012 Revision, GAO-12-208G (Washington, D.C.: Jan. 31, 2012).

[52] STEP was implemented by localities. Each locality hired and directed contractors to provide repair services to residents.

[53] Individuals may receive rental assistance through IHP after receiving STEP assistance in some circumstances. Because the authority for each program comes from different sections of the Stafford Act, the amount of STEP assistance individuals received did not count toward the $31,900 maximum available through IHP. There may be practical reasons to allow individuals to receive rental assistance after STEP services are provided. For instance, according to New York city officials, some individuals remained in hotels after repairs were finished because their homes still harbored mold, which was not addressed by STEP.

[54] 79 Fed. Reg. 16015 (Mar. 24, 2014).

[55] Our analysis excludes assistance to homeowners for flood damage to properties located in a special flood hazard area because such flood damage is typically not covered by private homeowners insurance. Also, individuals who received IHP assistance for home repair or personal property and reported having private insurance may have received assistance for damages not covered by their insurance, which is allowable.

[56] Certain provisions found in homeowners policies can also affect coverage. For example, while the most common private insurance policies typically exclude damage caused by flood and cover damage caused by wind, some insurance policies contain a clause that denies coverage for damage that is simultaneously caused by both an excluded peril (such as flood) and a covered peril (wind). According to insurance industry experts, this type of clause can present a particular challenge for owners of coastal properties when hurricanes occur. See GAO, Homeowers Insurance: Multiple Challenges Make Expanding Private Coverage Difficult, GAO-14-179 (Washington, D.C.: Jan. 30, 2014).

[57] GAO-06-954T.

[58] For more information on this example, see GAO's Government Data Sharing Community of Practice Minutes from the November 20, 2013 meeting. Minutes can be found at: http://www.gao.gov/assets

End Notes for Appendix I

[1] We chose to use data provided by state departments of correction in New York, New Jersey, and Connecticut because those states were most heavily impacted by Hurricane Sandy.

[2] PKEMRA was enacted as Title VI of the Department of Homeland Security Appropriations Act, 2007, Pub. L. No. 109-295, § 601, 120 Stat. 1355, 1394 (2006).

[3] GAO, Individual Disaster Assistance Programs: Framework for Fraud Prevention, Detection, and Prosecution, GAO-06-954T (Washington, D.C.: July 12, 2006); Federal Information System Controls Audit Manual (FISCAM), GAO-09-232G (Washington, D.C.: February 2009); and Standards for Internal Control in the Federal Government, GAO/AIMD-00-21.3.1 (Washington, D.C.: November 1999).

[4] Pub. L. No. 93-288, 88 Stat. 143 (1974) (codified as amended at 42 U.S.C. §§ 5121– 5207).

[5] GAO, Individual Disaster Assistance Programs: Framework for Fraud Prevention, Detection, and Prosecution, GAO-06-954T (Washington, D.C.: July 12, 2006); and Designing Evaluations: 2012 Revision, GAO-12-208G (Washington, D.C.: Jan. 31, 2012).

In: Hurricane Sandy Disaster Relief
Editor: Edward S. Ramirez
ISBN: 978-1-63482-174-2

Chapter 2

SMALL BUSINESS ADMINISTRATION: ADDITIONAL STEPS NEEDED TO HELP ENSURE MORE TIMELY DISASTER ASSISTANCE*

United States Government Accountability Office

WHY GAO DID THIS STUDY

On October 29, 2012, Hurricane Sandy made landfall, causing an estimated $65 billion in damage. SBA administers the Disaster Loan Program, which provides physical disaster loans (used to rebuild or replace damaged property) and economic injury disaster loans (used for working capital until normal operations resume) to help businesses and individual homeowners recover from disasters. In the aftermath of Hurricane Sandy, Congress passed the Disaster Relief Appropriations Act of 2013, which appropriated $779 million to SBA for disaster loans and administrative expenses.

GAO was asked to review SBA's assistance to small businesses following Hurricane Sandy. This report examines (1) the timeliness of SBA's disaster assistance to small businesses; (2) the loan approval rates for small businesses and reasons for decline for Hurricane Sandy and previous disasters; and (3) the

* This is an edited, reformatted and augmented version of a United States Government Accountability Office publication, No. GAO-14-760, dated September 2014.

extent to which SBA has implemented programs mandated by the Small Business Disaster Response and Loan Improvements Act of 2008. GAO analyzed SBA data on application processing; reviewed documentation related to SBA's planning, relevant legislation, and regulations; and interviewed SBA officials.

What GAO Recommends

GAO recommends that SBA revise its disaster planning documents, conduct a formal documented evaluation of lenders' feedback on IDAP, and report to Congress on challenges to implementing the program. SBA generally agreed with GAO's recommendations.

What GAO Found

Following Hurricane Sandy, the Small Business Administration (SBA) did not meet its timeliness goal for processing business loan applications. From application receipt to loan decision, SBA took an average of 45 days to process physical business disaster loans and 38 days for economic injury loans, both of which exceed SBA's 21-day application processing goal. SBA said it was challenged by an unexpectedly high volume of loan applications that it received early in its response to the disaster, in addition to other challenges, such as technological issues. SBA estimated that application submissions would peak about 7 to 9 weeks after Hurricane Sandy, but it received a larger volume of applications than were expected prior to that period. While SBA officials told GAO that the agency has taken steps to respond to some challenges, it has not revised its disaster planning documents—including the Disaster Preparedness and Recovery Plan—to reflect the early volume of application submissions received after Hurricane Sandy and the potential impact a similar experience could have on staffing, resources, and forecasting models for future disasters. Federal internal control standards state that management should identify risks and take action to manage them. Without taking its experience with early application submissions after Hurricane Sandy into account in its disaster planning documents and analyzing the potential risk early submissions may pose for timely disaster response, SBA may be unprepared for a large volume of applications to be submitted quickly following future disasters, which may result in delays in loan funds for disaster victims.

SBA approved 42 percent of business loan applications following Hurricane Sandy. This rate was lower than those of Hurricanes Katrina, Rita, and Wilma, higher than that of Ike, and comparable to that of Irene (the five disasters that generated the most SBA disaster loan applications since 2005). For Hurricane Sandy and for previous disasters, SBA declined business loan applications primarily because of applicants' lack of repayment ability and credit history.

SBA has not implemented the guaranteed disaster loan programs Congress mandated in 2008, including the Immediate Disaster Assistance Program (IDAP)—a bridge loan program through private sector lenders providing disaster victims with up to $25,000 with a 36-hour application approval period. SBA officials told GAO they are trying to implement IDAP but have received feedback from lenders that some program requirements—such as a statutory minimum 10- year time frame for servicing the loan under certain circumstances—may discourage lenders from participating. However, SBA has not conducted a formal documented evaluation of lenders' feedback that would establish the basis for proposed changes to requirements for Congress to consider. Without an appropriately documented evaluation of its outreach to lenders, SBA may not have complete and reliable information on which to base its reporting to Congress about its challenges with implementing the programs required by the act.

ABBREVIATIONS

CDBG	Community Development Block Grant
DCMS	Disaster Credit Management System
DPRP	Disaster Preparedness and Recovery Plan
EDALP	Expedited Disaster Assistance Loan Program
EIDL	economic injury disaster loan
FEMA	Federal Emergency Management Agency
IDAP	Immediate Disaster Assistance Program
IRS	Internal Revenue Service
LAA	Loan Authorization and Agreement
ODA	Office of Disaster Assistance
OIG	Office of Inspector General
PDAP	Private Disaster Assistance Program

PDC	Processing and Disbursement Center
SAPP	Sandy Alternative Processing Program
SBA	Small Business Administration
SBDC	Small Business Development Center

* * *

September 29, 2014

The Honorable Nydia M. Velázquez
Ranking Member
Committee on Small Business
House of Representatives

Dear Ms. Velázquez,

Hurricane Sandy made landfall along the southern New Jersey shore on October 29, 2012, causing historic devastation and substantial loss of life. Although 24 states were affected, New Jersey and New York (particularly the New York City metropolitan area) were the most impacted states—experiencing power outages and damage to public infrastructure, homes, and businesses. The National Oceanic and Atmospheric Administration estimates that damage from Hurricane Sandy totaled approximately $65 billion, making it the second costliest Atlantic storm since 1980 (after Hurricane Katrina in 2005). In the aftermath of Hurricane Sandy, Congress passed the Disaster Relief Appropriations Act of 2013, which appropriated $779 million to the Small Business Administration (SBA) for disaster loans and administrative expenses.[1]

While SBA is known primarily for its financial support of small businesses, the agency also plays a critical role in assisting the victims of natural and other declared disasters through its Disaster Loan Program. This program provides assistance to businesses of all sizes and homeowners affected by federally declared disasters. In the wake of Hurricane Sandy, concerns were raised regarding the timeliness of the financial assistance business owners received. In particular, there have been questions regarding the extent to which the program has been improved since Hurricane Katrina, and whether previously identified deficiencies, such as delays in processing disaster loan applications, have been addressed. Post-Katrina reforms have included increasing the capacity of SBA's electronic loan-processing platform and addressing requirements in the Small Business Disaster Response and Loan Improvements Act of 2008, such as updating SBA's disaster response plan.[2]

You asked us to examine SBA's response to Hurricane Sandy. Specifically, this report discusses (1) the timeliness of SBA's disaster assistance to small businesses following Hurricane Sandy and the factors that affected SBA's timeliness; (2) the loan approval rates for small businesses and reasons for application decline and withdrawal and loan cancellation following Hurricane Sandy, in comparison with those of previous disasters; and (3) the extent to which SBA has implemented loan programs mandated by the Small Business Disaster Response and Loan Improvements Act of 2008.

To evaluate the timeliness of SBA's disaster assistance to small businesses following Hurricane Sandy, we obtained and analyzed data on stages of the application process, such as application receipt to loan decision, from SBA's Disaster Credit Management System (DCMS), and interviewed SBA officials knowledgeable about these data. To identify factors that affected the timeliness of SBA's assistance, we reviewed documentation related to SBA's loan processing procedures and its internal assessment of its response to Hurricane Sandy, and we interviewed knowledgeable SBA officials. We also spoke with Small Business Development Centers and other local business organizations, such as chambers of commerce, to identify challenges businesses faced in receiving timely disaster assistance. To compare loan approval rates and describe the reasons why disaster business loan applications were declined or withdrawn and loans were cancelled following Hurricane Sandy and previous disasters, we obtained and analyzed DCMS data for Sandy and the five largest disasters since 2005 (as measured by application volume): Hurricanes Katrina, Rita, Wilma, Ike, and Irene. We also performed comparisons on the DCMS data we received to ensure its accuracy and completeness and interviewed SBA officials responsible for maintaining the DCMS data. We concluded that SBA's data were sufficiently reliable for the purposes of our report. To assess the extent to which SBA has implemented programs mandated by the Small Business Disaster Response and Loan Improvements Act of 2008, we reviewed relevant legislation and regulations, and interviewed officials knowledgeable about these matters. See appendix I for more detailed information on the scope and methodology of this report.

We conducted our work from August 2013 through September 2014 in accordance with generally accepted government auditing standards. Those standards require that we plan and perform the audit to obtain sufficient, appropriate evidence to provide a reasonable basis for our findings and conclusions based on our audit objectives. We believe that the evidence obtained provides a reasonable basis for our findings and conclusions based on our audit objectives.

BACKGROUND

SBA provides funding and assistance to individuals and businesses after disasters declared by either the President or SBA.[3] Administered by SBA's Office of Disaster Assistance (ODA), the Disaster Loan Program is the primary federal program for funding long-range recovery for nonfarm businesses that are victims of disasters and is the only form of SBA assistance not limited to small businesses. Disaster victims—including homeowners and renters, as well as businesses—initially register with the Federal Emergency Management Agency (FEMA) for assistance. FEMA automatically refers applicants whose income exceeds certain income thresholds to SBA's Disaster Loan Program.[4]

SBA's Disaster Preparedness and Recovery Plan (DPRP) outlines how the agency conducts operations in support of federal disaster response efforts. The plan is intended to ensure a broad scope of coordination, awareness, and support throughout the organization, and it describes how SBA conducts its disaster-related missions. The execution of SBA's disaster-related mission, as set forth in the DPRP, involves both response and recovery—including the recovery functions performed by SBA's Disaster Loan Program.[5]

The Small Business Act authorizes SBA to make available several types of disaster loans, including two types of direct loans: physical disaster loans and economic injury disaster loans.

- *Physical disaster loans:* These loans are for permanent rebuilding and replacement of uninsured or underinsured disaster-damaged property. They are available to homeowners, renters, businesses of all sizes, and nonprofit organizations. These loans are intended to repair or replace the disaster victims' damaged property to its predisaster condition up to a certain capped amount (see table 1). Almost any business concern or charitable or other nonprofit entity whose property is damaged in a declared disaster area is eligible to apply for a physical disaster loan; however, businesses in agriculture-related industries—also known as agricultural enterprises—are not eligible.[6]

Table 1. Characteristics of Selected SBA Disaster Loans Available to Businesses

Type of loan	Funds use	Eligibility	Collateral requirements[a]	Lending limit[b]	Interest rate[c]	Loan term
Physical disaster business loans	Repair or replace disaster-damaged property owned by the business, including real estate, inventories, supplies, machinery, and equipment.	Businesses of all sizes and private, nonprofit organizations such as charities, churches, and private universities.	Loans greater than $14,000 must be secured with collateral. For major disasters, loans greater than $25,000 must be secured with collateral.	$2 million aggregate for the repair or replacement of real estate, inventories, machinery, equipment, and all other physical losses.	Varies, but generally for those who cannot obtain credit elsewhere, interest is capped at 4 percent; for businesses that can obtain credit elsewhere, interest is capped at 8 percent.	Up to 30 years if no credit available elsewhere. For businesses with credit available elsewhere, maximum loan term is 7 years.
Economic injury disaster loans	Working capital loans to help meet ordinary and necessary financial obligations that cannot be met as a direct result of the disaster.	Small businesses, small agricultural cooperatives, and most private, nonprofit organizations of all sizes.	Loans greater than $25,000 must be secured with collateral.	$2 million aggregate for alleviating economic injury caused by the disaster.	Capped at 4 percent.	Not more than 30 years.

Source: GAO analysis of SBA documents. I GAO-14-760

[a] Loans for which borrowers pledge collateral are called "secured loans"; all others are referred to as "unsecured." SBA's preferred form of collateral is real estate. In an interim final rule published in the *Federal Register* on April 25, 2014, SBA raised the limits on unsecured loans to $25,000 for all EIDLs and for all physical disaster loans during major disasters. For disasters not declared major, the limit for unsecured physical disaster loans remains $14,000.

[b] The $2 million limit for business loans applies to the combination of physical and economic injury loans, and it applies to all disaster loans to a business and its affiliates for each disaster.

[c] Interest rates are periodically adjusted and vary for each disaster.

- *Economic injury disaster loans (EIDL):* These loans provide small businesses that are not able to obtain credit elsewhere with necessary working capital until normal operations resume after a disaster declaration.[7] The loans cover operating expenses the business could potentially have paid had the disaster not occurred. The Small Business Act restricts EIDLs to small businesses and private nonprofit organizations.[8]

Table 1 describes the characteristics of each of these two types of loans.

Not all businesses are eligible for both types of loans. Businesses of all sizes may apply for physical disaster loans, but only small businesses are eligible for an EIDL.[9] Small businesses that did not sustain physical damage from a disaster cannot apply for a physical disaster business loan but may still apply for an EIDL, and any small business that applies for a physical disaster business loan can also be considered for an EIDL. Such applicants may be approved for a physical disaster business loan but denied for an EIDL, or vice versa.[10]

Business owners may apply in person at a FEMA-established Disaster Recovery Center, at an SBA-established Disaster Loan Outreach Center or Business Recovery Center, by mail, or via SBA's Electronic Loan Application system. SBA requires applicants seeking either type of business disaster loan to submit a variety of documents as part of their application package, including the most recent federal income tax return, a signed copy of the Internal Revenue Service (IRS) tax information authorization Form 8821, a schedule of liabilities, and a current (dated within 90 days of the application) personal financial statement.[11] There are also additional filing requirements for applicants seeking an EIDL, such as monthly sales figures for the past 3 years and a request for a written explanation of the economic loss caused by the declared disaster.

SBA procedures generally require small businesses to submit an application for a physical disaster business loan no later than 60 days following the disaster declaration and no later than 9 months after this date for an EIDL.

12 SBA may authorize an extension of the filing period. Disaster declarations from Hurricane Sandy had different effective dates; therefore, deadlines for applying to the disaster loan program varied by state. For example, small businesses in both New York and New Jersey initially had until December 31, 2012, and July 31, 2013, to apply for physical disaster business loans and EIDLs, respectively.

However, each of these states was granted four separate extensions for physical disaster business loans (though none for EIDLs), and the final application deadlines for New York and New Jersey were April 13, 2013, and May 1, 2013, respectively.

SBA's regulations contain underwriting criteria that require, among other things, reasonable assurance of repayment, satisfactory credit, and satisfactory character.[13]

The regulations state that SBA must have reasonable assurance that all disaster loan applicants can repay their loans based on SBA's analysis of the applicants' credit or personal or business cash flow. The regulations also state that SBA is prohibited from lending to businesses that are engaged in lending or speculation or engaged in any illegal activity.[14]

SBA disaster loan application processing involves several stages (see fig. 1):

- *Application entry stage.* During the application entry stage, SBA screens all incoming applications to determine if they are acceptable.[15] Staff members at SBA's Processing and Disbursement Center (PDC) enter any application information submitted in paper format into the Disaster Credit Management System (DCMS), and where necessary, coordinate and consolidate any supplemental documentation received into one application per applicant.[16] SBA can also make automatic declines and pre-loss verification declines at this stage.[17]
- *Loss verification stage (physical disaster loans).* After the application entry stage, physical disaster loan applications move to the loss verification stage, while EIDL applications proceed directly to the application processing stage. During the loss verification stage, loss verifiers conduct on-site damage inspections for physical disaster loan applications to estimate the cost of restoring damaged property to predisaster condition. The verified loss becomes the basis for the loan amount.

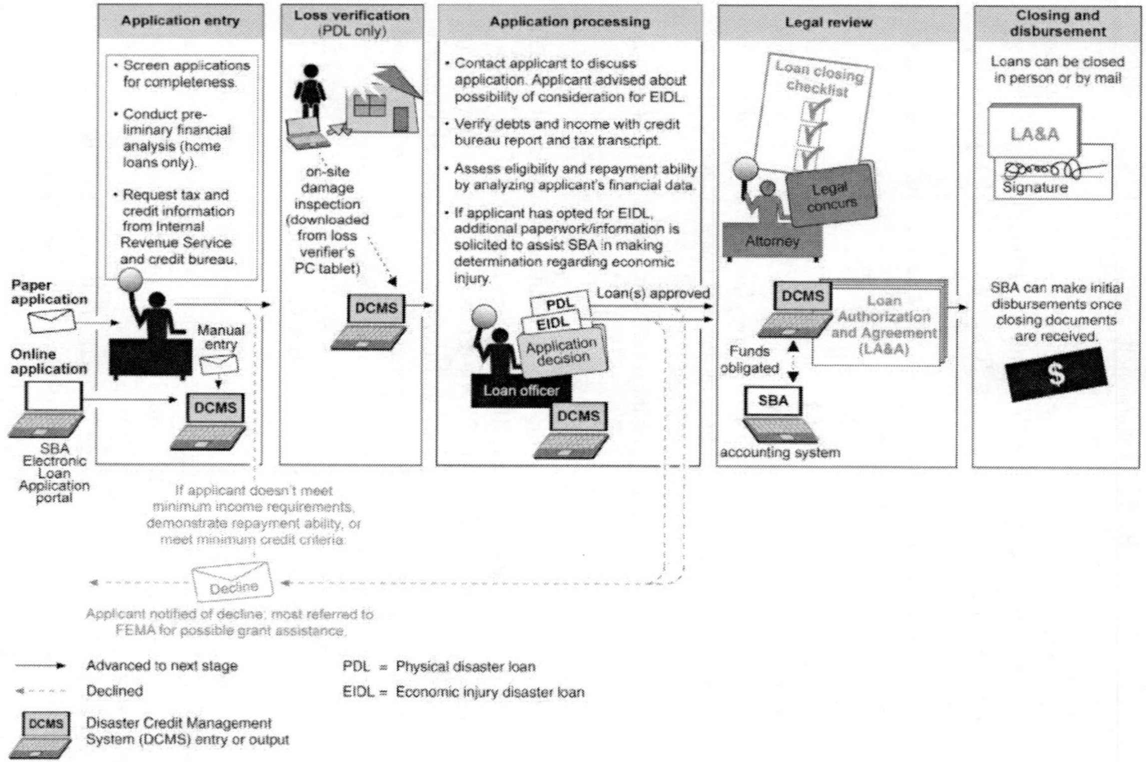

Source: GAO. | GAO-14-760.

Figure 1. SBA's Disaster Loan Process.

- *Application processing stage.* Once the loss verification is complete, an application moves to the application processing stage, where loan officers check for duplication of benefits and assess the applicant's credit history and ability to obtain credit elsewhere.[18] Loan officers also examine other applicant eligibility criteria, including compliance with child support obligations and history on other federal debt, such as student loans. Loan officers use a financial analysis tool within DCMS to determine if the applicant has the ability to repay the loan.[19]
- *Legal review stage.* For secured loans, legal staff members review the draft loan authorization and agreement for sufficiency of collateral instruments and other legal concerns.[20] They also create a loan-closing checklist—a list of the requirements necessary to generate the loan closing and other legal documents. Attorneys enter a legal concurrence into DCMS, which obligates the loan funds through an interface with SBA's accounting system. Legal support staff members prepare closing documents and mail them to the applicant or nearest Disaster Recovery Center.
- *Closing and disbursement.* For Hurricane Sandy loans, SBA could make a maximum initial disbursement, in the absence of collateral, of up to $14,000 for physical disaster business loans and $5,000 for EIDLs, once the agency received signed closing documents from the applicant. SBA could make a maximum initial disbursement of up to $50,000 for physical disaster loans and full disbursements for EIDLs, once all requirements were met. SBA generally makes subsequent disbursements on physical disaster loans based on the applicant's needs and how they spent prior disbursements. SBA procedures also generally require small businesses to arrange for and obtain all loan funds within 6 months from the date of the loan authorization and agreement.

Following Hurricane Sandy, SBA received disaster loan applications from 9 states and Puerto Rico, with the majority of the 15,745 business disaster loan applications generated by businesses in New Jersey and New York.[21] See figure 2 below for a map of highly impacted counties in New Jersey and New York and information on business disaster loan applications from these counties. (See Appendix II for additional information regarding Figure 2.)

Congress enacted the Small Business Disaster Response and Loan Improvements Act of 2008 to expand steps taken by SBA after Hurricane Katrina and require new measures to help ensure that SBA would be prepared for future disasters. The act included three provisions requiring SBA to issue regulations to establish new guaranteed disaster programs using private sector lenders:[22]

- **Expedited Disaster Assistance Loan Program (EDALP)**. EDALP would provide small businesses with expedited access to short-term guaranteed loans of up to $150,000.
- **Immediate Disaster Assistance Program (IDAP)**. IDAP would provide small businesses with interim guaranteed loans of up to $25,000 with a loan decision within 36 hours.
- **Private Disaster Assistance Program (PDAP)**. PDAP would make guaranteed loans available to homeowners and small businesses in an amount up to $2 million.

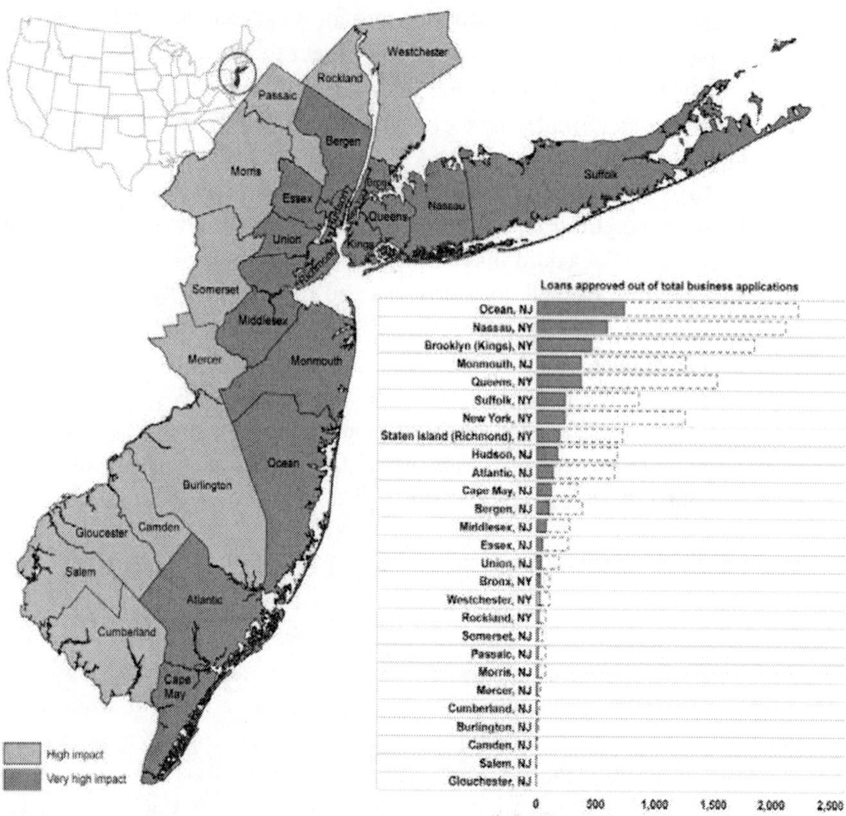

Source: GAO analysis of FEMA and SBA data; MapInfo (map). I GAO-14-760.

Note: This map does not include all counties that were impacted by Hurricane Sandy and all counties that received SBA business disaster loans. The Federal Emergency Management Agency (FEMA) defines "Very high" impact areas as those in which more than 10,000 of the county's population were exposed to the storm's surge. "High" impact areas are those in which 500 to 10,000 of the county's population were exposed to the storm's surge, modeled wind damages was greater than $100 million, or precipitation was greater than 8 inches. In addition to applications that were approved and declined, "Total applications received" includes applications that were ultimately withdrawn, which are not factored into SBA's calculation of approval rates. A table containing the data presented in this graphic can be found in Appendix II.

Figure 2. New Jersey and New York Hurricane Sandy Business Disaster Loan Applications Received, Approved, and Dollar Amount (as of January 2014).

See table 2 for characteristics of these loan programs.

Table 2. Characteristics of Guaranteed Loan Programs Mandated by the Small Business Disaster Response and Loan Improvements Act of 2008

IDAP		EDALP	PDAP
Purpose	To provide eligible small business concerns with emergency, small-dollar guaranteed loans within 36 hours after SBA receives an application and pending the applicant's receipt of a conventional disaster loan.	To provide guaranteed bridge loans to disaster victims eligible for the 7(b) disaster assistance program that need a greater amount of funding. The loans are also intended to be disbursed more quickly than a standard SBA loan.	To provide guaranteed loans made by qualified lenders to eligible small businesses (and homeowners).
Disaster declaration	"Major," "SBA Administrative Disaster" or "SBA EIDL Disaster"	"Catastrophic"	"Catastrophic"
Loan amount	Up to $25,000	Up to $150,000	Up to $2 million
Loan terms and conditions	• Decision within 36 hours • Made to small businesses • SBA guaranteed share equals 85 percent • Repaid from disaster loan funds if borrower's SBA disaster loan is approved • During initial period borrower will pay interest	• Made to small businesses • SBA may guarantee the timely payment of principal and interest to lender • Decision: Loan may receive expedited loss verification and loan	• Made to small businesses and homeowners • SBA may guarantee up to 85 percent of any loan • SBA may guarantee the timely payment of principal and interest to

Table 2. (Continued)

IDAP		EDALP	PDAP
	only on disbursed principal balance of IDAP loan • Borrower has at least 10 years from the date of final disbursement of the IDAP loan but no more than 25 years from the date of the final disbursement to repay the loan[a] • Initial Period: Begins upon the initial disbursement of an IDAP loan and ends upon: full repayment of the IDAP loan from the proceeds of the IDAP borrower's disaster loan; SBA notice to the IDAP lender of decline of the IDAP borrower's Disaster Loan application; or receipt by the IDAP lender of partial repayment of the IDAP loan from the proceeds of the disaster loan • Term Period: The repayment period begins following: SBA notice to the IDAP lender of decline of the IDAP Borrower's disaster loan application; receipt by the IDAP lender of partial repayment of the IDAP loan from the proceeds of the disaster loan; or final disbursement of the IDAP loan, whichever is later, and ends when the IDAP loan is repaid in full.	processing if borrower is a major source of employment in the disaster area or vital to recovery efforts in the region • Not to exceed 180 days[b]	lender • Up to 30 years[c]

Source: GAO analysis of the Small Business Response and Loan Improvements Act of 2008 and related SBA regulations. I GAO-14-760

[a]Only if SBA declines the IDAP borrower's disaster loan application or the approved amount is insufficient to repay the IDAP loan in full.

[b]SBA Administrator may extend on a case-by-case basis.

[c]For economic injury disaster loans and physical home disaster loans, borrowers have up to 30 years to repay the loan. For physical business disaster loans, borrowers have up to 30 years to repay the loan if they do not have credit elsewhere. Otherwise, the borrower has 7 years to repay the loan.

UNANTICIPATED VOLUME OF EARLY APPLICATION SUBMISSIONS AND DELAYED INCREASE IN STAFF HINDERED SBA'S ABILITY TO PROVIDE TIMELY ASSISTANCE

Following Hurricane Sandy, SBA did not meet its timeliness goal of processing business loan applications from receipt to loan decision within 21 days for a number of reasons, and a backlog of applications developed rapidly. SBA officials said the agency was challenged by a large, unanticipated volume of applications early in its response to the disaster. In addition, SBA relied on inaccurate estimates of how quickly staff could process applications, which delayed its decision to increase the numbers of staff to process the applications. SBA also faced challenges related to its information system, among others. According to SBA, the agency is taking steps to address some of the challenges it faced after Hurricane Sandy. However, SBA has not revised its disaster planning to reflect the unanticipated volume of early applications it received. As a result, SBA risks continuing to be unprepared for a large number of disaster loan applications to be submitted at the beginning of a disaster response.

SBA Did Not Meet Its Timeliness Goal for Application Processing and a Backlog of Applications Grew Rapidly

Following Hurricane Sandy, SBA did not meet its goal to process business loan applications within 21 days from application receipt to loan decision.[23] As shown in figure 3, SBA took an average of 45 days for physical disaster business loan applications and 38 days for EIDL applications.[24]

The average processing time for disaster business loans peaked in March 2013 (5 months after the storm); business loans for which SBA reached a decision in March 2013 had been in processing for an average of nearly 60 days. One year after the storm, SBA's processing times for business loan applications still exceeded its goal of 21 days.

Following Hurricane Sandy, SBA met its goal to provide first loan disbursements to 95 percent of borrowers within 5 days of loan closing. On average, SBA provided first disbursements to approved businesses within 5 days of closing. According to SBA officials, loan funds are generally provided within four disbursements over a period of time after closing.

While there are no timeliness goals associated with closing a loan, borrowers who have been approved for a disaster loan have 60 calendar days from the date of the loan authorization agreement to close it. In order to close the loan, borrowers must sign and return all documents required for an initial disbursement, which is then reviewed by SBA's Chief Legal Advisor. For Hurricane Sandy, it took on average an additional 66 days from approval to close a physical disaster business loan and an additional 43 days for an EIDL.

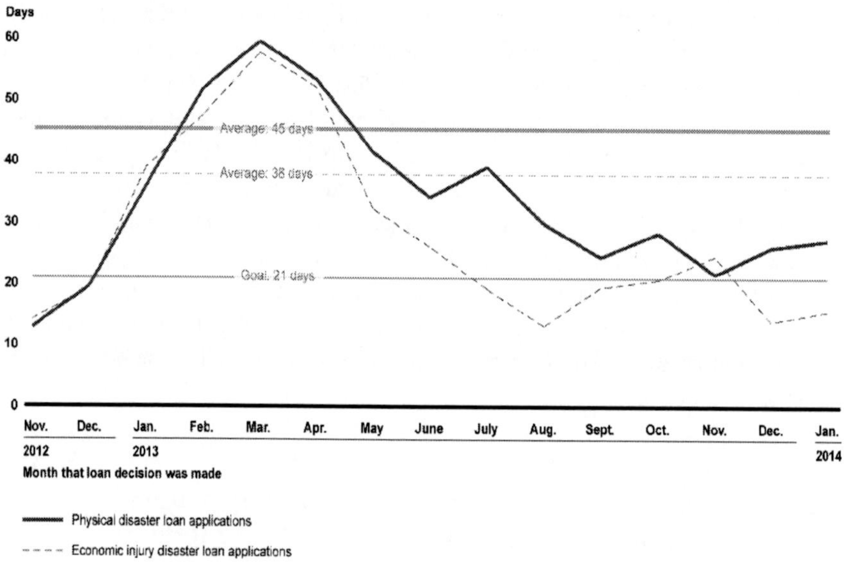

Source: GAO analysis of Small Business Administration data. | GAO-14-760.

Figure 3. Average Days from Receipt to Loan Decision for Hurricane Sandy Business Disaster Loans by Loan Type, November 2012-January 2014.

A backlog of applications that were "in processing" (meaning those SBA had received but for which it had not yet made a loan decision) grew rapidly over the course of SBA's response to the disaster. Figure 4 illustrates the backlog of applications that developed over the course of SBA's response to Hurricane Sandy. As of January 2014, SBA officials noted that five home applications, five physical disaster loan applications, and two EIDL applications remained in processing.[25]

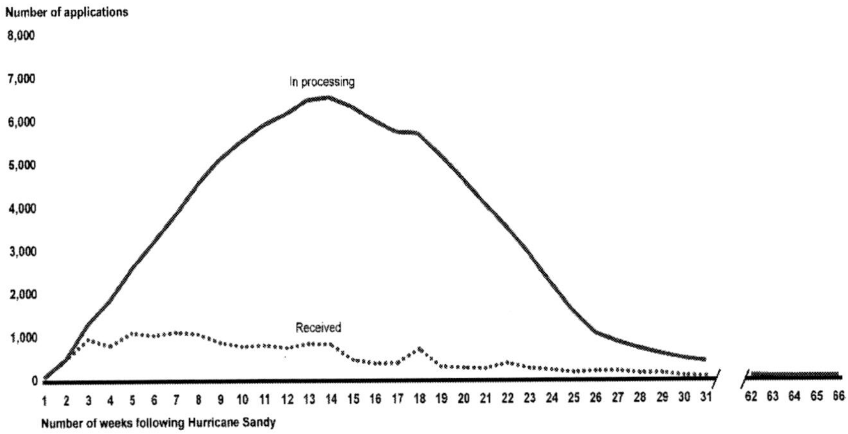

Source: GAO analysis of Small Business Administration data. | GAO-14-760.

Note: The figure depicts the number of weeks from Hurricane Sandy's landfall on October 29, 2012. The final week spans from January 27, 2014, to February 2, 2014. Business applications "received" represent the number of business disaster loan applications that were submitted to SBA during the week. Business applications "in processing" is the cumulative total of applications on which SBA had not yet made a loan decision (approval or decline) and represents the backlog for business disaster loan applications.

Figure 4. Number of Hurricane Sandy Business Disaster Loan Applications Received and in Processing, October 2012- January 2014.

SBA Did Not Anticipate Receiving a High Volume of Loan Applications Early in Its Response to Hurricane Sandy

SBA said that in the aftermath of Hurricane Sandy, it was challenged by a high volume of loan applications submitted at a faster rate than it had experienced in previous disasters. SBA's initial estimates of when it would receive applications differed from when it actually did receive them. To prepare for a disaster, SBA uses assumptions about the volume and timing of the applications it expects to receive based on historical data— known as the "application intake curve." SBA inputs these assumptions into forecasting models that predict the staff levels necessary to meet processing needs. According to the application intake curve for Hurricane Sandy, SBA estimated that application submission would peak about 7 to 9 weeks after Hurricane

Sandy. However, as shown in figure 5, SBA began receiving business applications earlier.

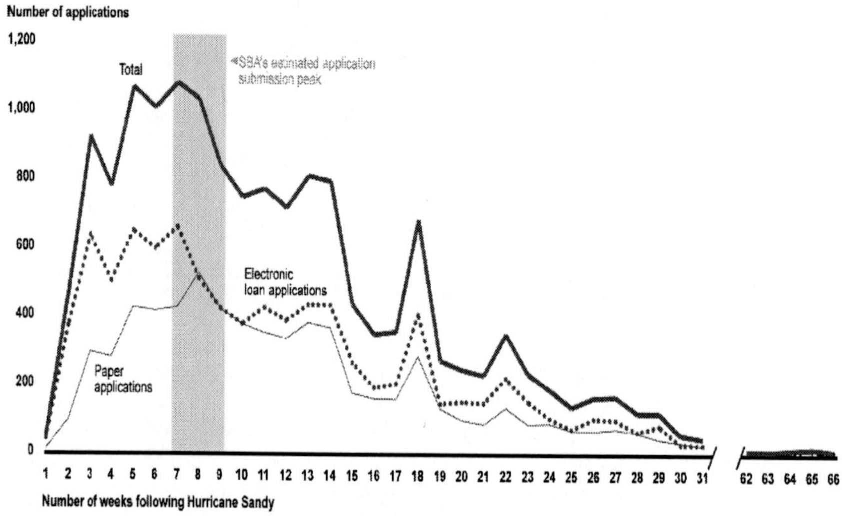

Source: GAO analysis of Small Business Administration data. | GAO-14-760.

Note: The graphic depicts the number of weeks from Hurricane Sandy's landfall on October 29, 2012. The final week spans from January 27, 2014 to February 2, 2014.

Figure 5. Number of Hurricane Sandy Business Disaster Loan Applications SBA Received from October 2012 through January 2014.

According to SBA, the early spike in applications occurred because a majority of applications were submitted electronically rather than on paper, which resulted in a large volume of applications within a few days of the disaster. SBA stated that the earlier receipt of electronic submissions was caused by the convenience and speed of the Internet-based application as well as the elimination of postal handling time. While SBA created electronic loan applications to simplify and expedite the application process and continues to encourage submitting applications electronically, SBA noted that it did not anticipate receiving such a large volume of electronic loan applications early in its response following Hurricane Sandy.

Based on its experience in fiscal year 2012, SBA initially estimated that it would receive between 11,000 and 21,600 business disaster loan applications after Hurricane Sandy and 36 percent of all disaster applications would be submitted electronically. Following Hurricane Sandy, SBA received 15,745 business disaster loan applications, and 55 percent of all disaster applications were submitted electronically. SBA also initially estimated that it would receive a certain percentage of each business loan type electronically, and these estimates were lower than the actual submissions (as shown in table 3).

Table 3. Estimated and Actual Percentages of Hurricane Sandy Disaster Loan Applications Submitted Electronically, Total and Business

	Total disaster loan applications[a]	Physical loan applications	EIDL applications
SBA estimate of electronic application submission	36%	32%	52%
Actual electronic application submission	55%	55%	72%

Source: GAO analysis of SBA data. | GAO-14-760

[a]Total disaster applications include all home and business disaster loan applications received following Hurricane Sandy. Home loan applications comprised about 80 percent of the total number of Hurricane Sandy disaster applications submitted to SBA.

In addition, SBA was unprepared for the use of multiple entry points through which applications could be submitted electronically. According to SBA officials, applicants could submit application documents electronically in a piecemeal fashion via numerous portals, including computer stations set up at each Disaster Loan Outreach Center and personal computers. In contrast, SBA noted that paper applications would only be accepted if all forms and documentation were submitted to the Processing and Disbursement Center (PDC). Because of the various entry points for electronic applications, SBA officials said that the PDC experienced difficulties in consolidating the application documents, which caused delays in processing the electronic applications.

According to SBA, the agency is taking steps to process electronic application submissions more effectively for future disasters based on its experience with Hurricane Sandy. Specifically, SBA officials noted that the agency is consolidating the number of entry points through which supporting documentation can be submitted electronically and is improving DCMS to allow documents from SBA follow-up requests to now be submitted electronically instead of in paper form. In addition, SBA officials noted that the agency expects to build an electronic portal where applicants can check the status of their applications and whether documents have been received. SBA officials noted that these efforts began in October 2013 and said they will be monitoring their effectiveness as they continue to automate and streamline the consolidation process.

Although SBA is making technological improvements, it has not updated its key disaster planning documents—namely the DPRP and *ODA Playbook*—to reflect the early volume of application submissions received after Hurricane Sandy and the potential impact that a similar experience could have on staffing, resources, and forecasting models for future disasters.[26] Federal internal control standards state that management should identify risk and that risk identification methods may include, among others, forecasting and strategic planning. Once risks have been identified, they should be analyzed for their possible effect. Risk analysis generally includes estimating the risk's significance, assessing the likelihood of its occurrence, and deciding how to manage the risk and what actions should be taken.[27] Further, according to the DPRP, the primary goals of forecasting and modeling are to predict as accurately as possible the application volume that will result from a disaster and the timing of when applications will be received. Without taking its experience with early application submissions after Hurricane Sandy into account in its plans, SBA risks continuing to be unprepared for a larger number of disaster applications to be submitted at the beginning of a future disaster response, potentially resulting in delays in receipt of loan funds for disaster victims.

Inaccurate Loan Processing Expectations Delayed the Decision to Increase Staff

One factor that affected the timeliness of SBA's disaster assistance was inaccurate expectations for the rate at which SBA staff could process loan applications, which caused SBA to delay its decision to increase staff levels. ODA officials said that the PDC communicated inaccurate production estimates to ODA headquarters, which led to delays in increasing staff levels to respond to the early influx of applications. According to ODA officials, management at PDC and ODA participated in daily conference calls during Hurricane Sandy to discuss, among other things, production levels of loan officers and any need to increase staff. ODA officials said that PDC management projected an average productivity level of 3 home loan applications a day per loan officer and 1.5 business loan applications a day per loan officer, for a combined average of 2.25 disaster loan applications a day. However, this productivity expectation was not met over the course of the response. According to ODA officials, the PDC loan officers were actually completing a combined average of 1.5 loan applications a day per loan officer, yet continued to maintain in communications with ODA that they could meet their initial productivity expectations. Because these estimates were based on production benchmarks established after Hurricane Katrina, ODA officials noted that they relied on the PDC's production estimates and delayed their decision to increase staff. However, ODA officials said they later recognized that the past disaster production rate was not an appropriate indicator of production following Hurricane Sandy due to, among other things, differences in the types of businesses impacted and the larger number of approved applications.

As shown in figure 6, although the number of business applications received peaked in December 2012, ODA ultimately added loan officers at its Sacramento loan center in mid-December 2012 and its Buffalo center in the beginning of March 2013, past the peak months of business application receipt.[28]

Per the DPRP, SBA has the flexibility to incrementally add pre-identified staff to its workforce in order to tailor its response to the size of the disaster and number of loan applications expected.[29] According to SBA, during the peak of staffing in March 2013, about a fifth of the loan processing staff belonged to its core or permanent staff; the remaining 80 percent were either pre-identified processing staff or new hires. Further, ODA officials noted that most of the additional processing staff members, particularly those in the Sacramento center, were new hires because it was more cost-effective to hire loan officers from the area than to incur costs of transporting pre-identified staff from various locations. As a result, SBA's decision to increase its workforce to respond to the growing backlog was further delayed.

Based on its experience with Hurricane Sandy, ODA told us that several changes have been made with regard to communication with the PDC and staffing increases. The PDC is now required to produce a new series of daily reports for ODA headquarters to increase transparency and improve communication during future disasters. Specifically, these reports include more detailed information on the PDC's production rates, number of applications submitted, and size of the application backlog. ODA has also created a standard template for centers to request and justify additional staff to ensure that they provide consistent information about, among other things, current and expected performance. Further, SBA is determining if adding permanent loan processing staff in centers other than the PDC, such as those in Sacramento and Buffalo, is needed to respond to disasters.

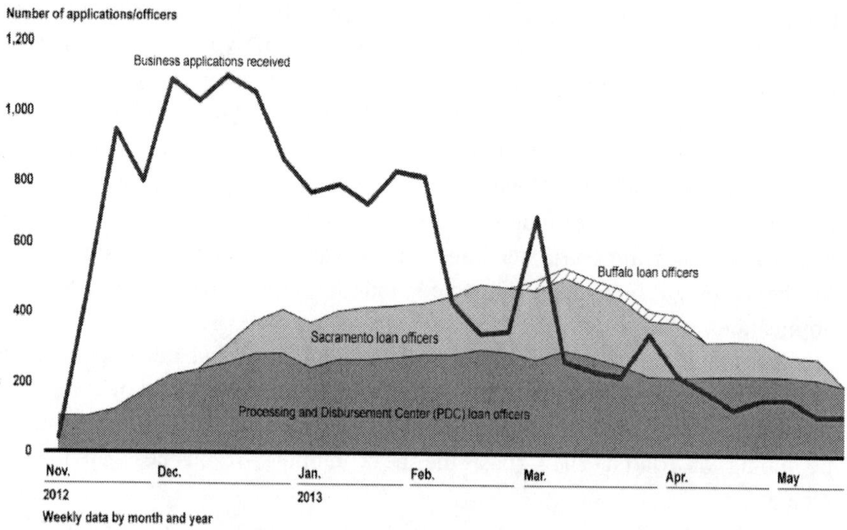

Source: GAO analysis of Small Business Administration data. | GAO-14-760.

Note: The weekly data for loan officers and business applications received differ by 1 day. For example, the final week of data for loan officers ends on May 18, 2013, and the final week of data for business applications ends on May 19, 2013.

Figure 6. Number of Active Loan Officers and Hurricane Sandy Business Disaster Loan Applications Received by Week, October 2012-May 2013.

SBA Faced Other Challenges to Processing Disaster Loan Applications and Has Taken Actions to Address Them

To address challenges with providing timely assistance following Hurricane Sandy, SBA made various changes to its loan processing approach, DCMS, and loan officer training. However, because SBA has not received a large volume of applications since Hurricane Sandy, it is too early to determine whether these changes will improve the timeliness of SBA's response for future disasters similar to Hurricane Sandy's magnitude.

- **Loan Processing Approach:** Following Hurricane Sandy, SBA processed loans in the order in which they were received, regardless of whether the applicant was a business or homeowner—referred to as a "first-in, first-out" approach. After Hurricane Sandy, SBA received over four times as many home loan applications as business applications, and these home loan applications were received earlier.[30] As a result, business owners could have faced delays due to the large number of home loan applications submitted ahead of them. In August 2013, the Hurricane Sandy Rebuilding Task Force recommended that SBA create two separate application tracks for home and business disaster loans, which SBA implemented in October 2013.[31] According to SBA officials, the agency had processed approximately 2,166 business disaster applications using the separate application tracks in an average of 12 days as of June 2014.
- **DCMS Challenges:** Over the course of its response, SBA encountered various challenges with DCMS, including server hardware crashes and periods of system latency (periods of slowness and freezing), which added to some delays faced by business owners in receiving disaster assistance. According to SBA, the agency is taking several steps to improve DCMS for future disasters. For example, SBA planned to institute a process for updating system equipment based on a consistent life-cycle by, among other things, conducting a baseline inventory and implementing a plan to replace outdated hardware. According to SBA officials, the inventory has been validated by center directors and the plan has been completed. In addition, SBA officials said that the agency has made improvements to its DCMS Help Desk, which responds to loan officers who experience system issues. These improvements include implementing a tiered approach to addressing system issues and better ensuring that the Help Desk is used by all centers in a standardized way.
- **Loan Officer Training:** As mentioned previously, most of the additional processing staff, particularly those at the Sacramento center, were new hires, but SBA found that these new loan officers were not effectively trained to quickly respond to the backlog of business applications. According to SBA, loan processing production was not fully realized immediately following the increase in staff because it took new loan officers longer than expected to fully learn how to process loan applications. In addition, loan officers were not converted into business loan officers until they demonstrated

proficiency in processing home loans. Further, these loan officers required additional specialized training to learn how to process a business loan within DCMS. According to SBA, this strategy was ineffective because it was implemented at the peak of receiving and processing applications. SBA has taken steps to revise its loan officer training for future disasters to be more production-oriented rather than focusing on DCMS use. According to SBA officials, the PDC developed a revised training course for processing business loans in May 2014 that is required for all loan officers. As of July 2014, three classes of approximately 18 loan officers each have completed the training. Additionally, SBA officials noted that it has reorganized its loan officers into two groups that specialize in processing home and business loans based on the previously mentioned changes made to its loan processing approach.

Small Business Development Centers and Other Local Organizations Identified Challenges Businesses Faced in Receiving Timely Disaster Assistance

Select Small Business Development Centers and local business organizations in New York and New Jersey that we met with identified two main challenges that affected the timeliness of SBA's assistance from the perspective of small businesses applying for loans: time-consuming loan documentation requirements and lack of SBA follow-up. According to SBA officials, the agency is taking actions to address these two challenges for future disasters.

Loan Documentation Requirements

Nearly all 14 SBDCs and local business organizations we met with noted that meeting the application's documentation requirements was time-consuming and onerous to business owners. For example, 11 of the 14 entities we met with told us that gathering the necessary documentation to complete an application was difficult given the physical damage caused by Hurricane Sandy. Further, 6 of the 14 entities noted that follow-up requests for additional or updated documentation by SBA prolonged the application process and loan decision.

As discussed previously, SBA requires business applicants to submit a variety of documents in order to make a loan decision, including the most recent federal income tax returns, a signed copy of the IRS tax information authorization Form 8821, a current (dated within 90 days of the application) personal financial statement, and a schedule of liabilities, among others. According to SBA, additional requests for documentation usually occur when either information needed to make a decision is missing or information provided was not sufficient. For example, many business owners submit their most recent federal income tax returns, but do not include this information for their affiliates, which is required to make a loan decision. In addition, tax returns are often requested—in addition to the IRS Form 8821, which provides SBA access to an applicant's previous tax transcripts—because they provide loan officers with more detailed information that may be needed to make a loan decision.

SBA officials said that the agency is taking several steps to streamline the documentation requirements for applicants and improve the process to submit additional information. Specifically, SBA has examined the entire loan application process in order to identify and eliminate documents that do not help loan officers make a decision on an application. According to SBA officials, the proposed changes to the required documentation have been drafted and will be incorporated in the next iteration of SBA's disaster loan program standard operating procedures by the end of 2014. Also, as previously mentioned, SBA officials told us that the agency plans to consolidate entry points through which electronic loan applications can be submitted and improve DCMS so that applicants can submit certain documents from follow-up requests electronically rather than sending paper copies in the mail.

Further, SBA took steps to reduce documentation requirements for applicants with strong credit scores by amending regulations to allow the agency to rely on credit scoring rather than cash flow when determining an applicant's ability to repay.[32] Under these new regulations, SBA has the option to base its repayment ability determination on either the applicant's cash flow or credit score. According to SBA, by removing the requirement to analyze cash flow for applicants with strong credit scores in order to determine ability to repay, the agency can process these applications more expeditiously. Further, SBA stated that applicants with strong credit scores will not be required to submit further information documenting their personal and business cash flows to SBA. Finally, SBA noted that credit scoring can help SBA dedicate staff to applicants that do not have strong credit and reduce the overall processing time for loan applications. SBA pursued these legal changes to determining repayment ability in response to recommendations made by the Hurricane Sandy Rebuilding Task Force as well as its experience with the

Sandy Alternative Processing Program (SAPP), in which selected home loan applications were processed in a similar manner after Hurricane Sandy.[33]

Applicant Follow-Up

Over half of the entities we met with said that business owners noted that there was a lack of SBA contact after submitting their applications, and many owners were unaware of the status of their application throughout the process, including whether or not it had been received at the PDC. Additionally, five of the entities noted that there was a lack of continuity with loan officers or case managers over the course of the application process.[34] Two of these five entities said that some business owners had experienced up to eight different loan officers or case managers during the process. In addition, these five entities reported that submitted documentation and information were lost when loan officers and case managers changed.

According to SBA officials, due to the physical damage caused by Hurricane Sandy, it was difficult for loan officers and case managers to contact applicants via telephone or e-mail despite their efforts. Further, because of the backlog of applications, SBA officials said that it took approximately 10 to 15 days for a loan officer to examine a submitted application and contact the borrower. SBA officials told us that contact with an applicant during specific points in the application process can be tracked within DCMS. According to SBA, one loan officer is generally assigned to an applicant until a loan decision is made; if the loan is approved, one case manager is typically assigned until the loan is fully disbursed. SBA officials told us that an applicant may have more than one loan officer or case manager for several reasons, including if the number of applications increases or if current loan officers or case managers are required to supervise newer staff. In this case, the PDC will notify the applicant of the change in loan officer or case manager as soon as possible.SBA officials told us that some documents can be misplaced during the application process due to the multiple ways applicants can submit information to the PDC. In addition, some documents may not be misplaced; rather, they may not yet have been entered into DCMS and may not be available for loan officers to view.

According to SBA officials, efforts to process electronic application submissions more effectively will address these issues. As part of this project, as previously mentioned, officials said SBA expects to create an electronic portal that will share information with applicants on the status of their applications and documents received, which should increase transparency and communication during the loan application process.

Approval Rates, Reasons for Decline, Withdrawals, and Cancellations for Hurricane Sandy and Previous Disasters

SBA's overall approval rate for Hurricane Sandy business loan applications was 42 percent, which was lower than those for Hurricanes Katrina, Rita, and Wilma, higher than that for Hurricane Ike, and comparable to the approval rate for Hurricane Irene. For Hurricane Sandy and each of the five previous disasters we examined, lack of repayment ability and unsatisfactory credit history were the two primary reasons why SBA declined business loan applications. Application withdrawal rates and loan cancellation rates were both higher for Hurricane Sandy than for the five previous disasters we examined.

Business Loan Approval Rates for Hurricane Sandy Were Higher Than Those of Some Previous Disasters but Lower Than Others

In comparison to the five previous disasters we selected, the overall approval rate for business loan applications for Hurricane Sandy was lower than those for Hurricanes Katrina, Rita, and Wilma, higher than that for Hurricane Ike, and comparable to the approval rate for Hurricane Irene.[35] The overall approval rate for Hurricane Sandy business loan applications, including both physical disaster loans and EIDLs, was 42 percent.[36] The approval rate for physical disaster loan applications was 45 percent, while the approval rate for EIDLs was 28 percent. For physical disaster loan applications specifically, the approval rate for Hurricane Sandy was lower than those for Hurricanes Katrina and Wilma, but higher than those for Hurricanes Ike and Irene, and comparable to that for Hurricane Rita. The approval rate for EIDL applications resulting from Hurricane Sandy was lower than those for Hurricanes Katrina, Rita, Wilma, and Irene, but higher than the rate for Ike (see fig. 7).

Hurricane Sandy resulted in the highest total approval rate of disaster loan applications in comparison to the five previous disasters we examined. The total approval rate for disaster loan applications includes home loan applications combined with physical disaster loan and EIDL applications.[37] For Hurricane Sandy, the total approval rate was 53 percent.[38]

Applicants' Repayment Ability and Creditworthiness Were the Primary Reasons Loan Applications Were Declined for Hurricane Sandy and Previous Disasters

Lack of repayment ability and unsatisfactory credit history were the two primary reasons why SBA declined business loan applications following Hurricane Sandy and in each of the five previous disasters we examined. Following Hurricane Sandy, SBA received 14,938 business loan applications (excluding reconsiderations and appeals), and declined 5,663 applications as of January 31, 2014. Of these declined applications, lack of repayment ability was at least one of the reasons SBA cited on 2,644 applications (47 percent), while unsatisfactory credit history was at least one of the reasons SBA cited on 2,317 applications (41 percent).[39] See figure 8 below for more information on the leading decline codes for Hurricane Sandy and previous disasters. Other reasons that Hurricane Sandy business loan applications were declined include unsatisfactory history on an existing or previous SBA loan or other federal obligation, unsatisfactory debt payment history, a finding that the business was not eligible due to recoveries from other sources, and availability of credit elsewhere, among others.

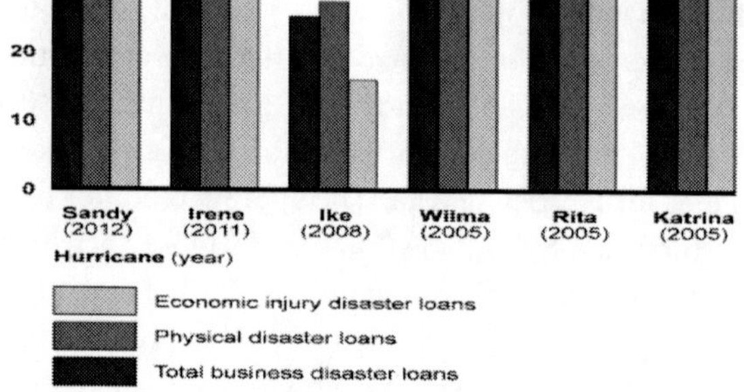

Source: GAO analysis of Small Business Administration data. | GAO-14-760.

Figure 7. Comparison of Business Application Approval Rates by Loan Type for Hurricane Sandy Compared with Previous Selected Disasters, 2005-2013.

For Hurricane Sandy, Application Withdrawal Rates and Loan Cancellation Rates Were Higher Than for Previous Disasters

Of the 14,558 original Hurricane Sandy business loan applications that had reached a decision status by January 31, 2014, 4,715 (approximately 32 percent) were withdrawn by either SBA or the applicant—a withdrawal rate higher than those of the previous disasters we examined.[40] The withdrawal rates for these previous disasters ranged from approximately 18 percent (for Hurricane Ike) to approximately 23 percent (for Hurricanes Katrina and Wilma), as shown in figure 9.

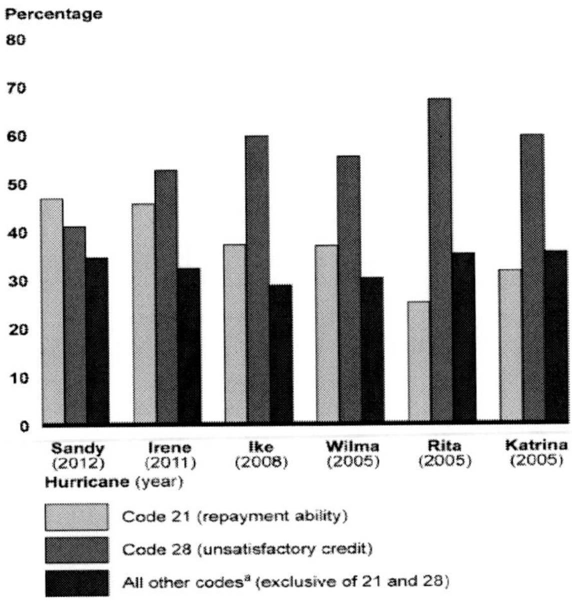

Source: GAO analysis of Small Business Administration data. | GAO-14-760.

Note: Percentages for each disaster sum to greater than 100 because applications can be assigned multiple decline codes.

[a]See appendix III for a description of the codes included in the "All others" category.

Figure 8. Frequency of SBA Decline Codes for Business Disaster Loan Applications for Sandy and for Previous Selected Disasters.

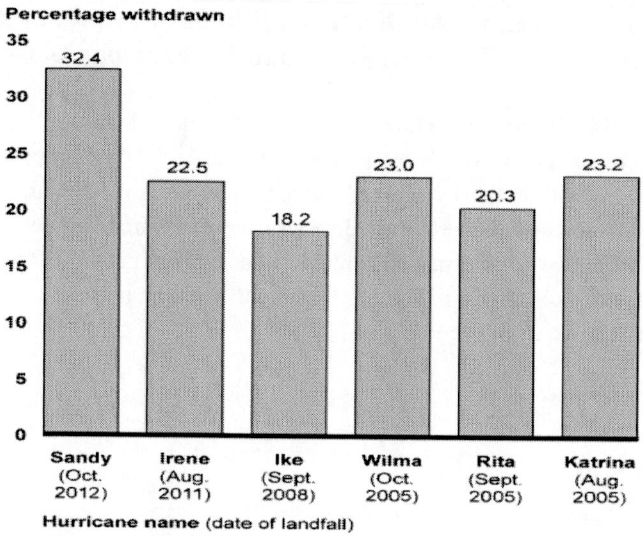

Source: GAO analysis of Small Business Administration data. | GAO-14-760.

Figure 9. Comparison of Business Application Withdrawal Rates for Hurricane Sandy Compared with Previous Selected Disasters, 2005-2013.

For Hurricane Sandy, approximately 60 percent of withdrawn applications were withdrawn by SBA, while the remaining 40 percent were withdrawn at the applicant's request. The percentage of SBA-initiated withdrawals was higher for two of the disasters we examined and lower for the others. SBA loan officers assign codes to withdrawn applications, indicating the reasons for withdrawal. Of the 14 withdrawal codes, 5 codes represent cases that originate from an applicant's request to withdraw the application from consideration, while 9 codes represent SBA's decision to withdraw the application.[41] The leading reason for withdrawal was the applicant's failure to provide SBA with all requested information, which resulted in 1,542 withdrawals (approximately 33 percent of all withdrawn applications).

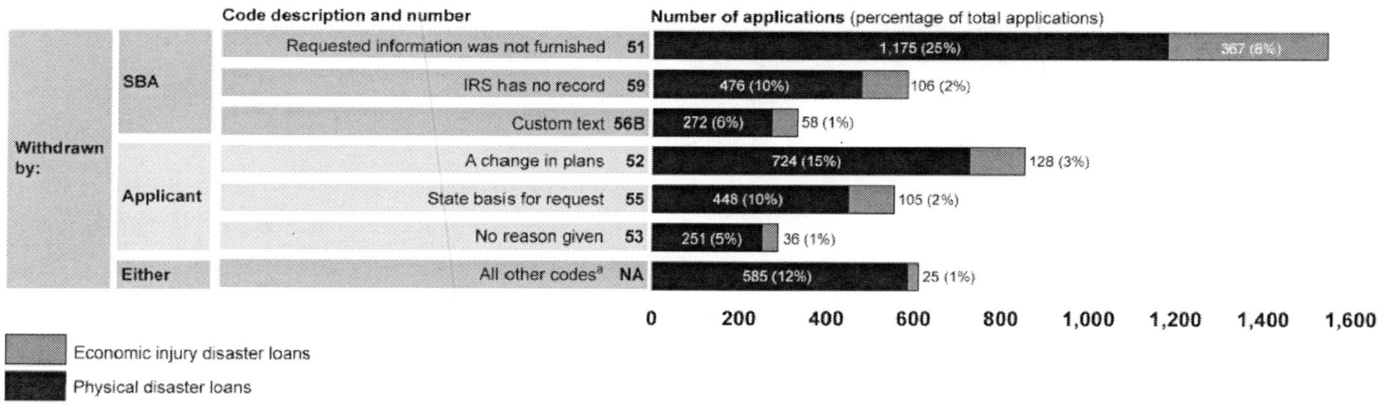

Source: GAO analysis of Small Business Administration data. | GAO-14-760.

Note: Percentages sum to greater than 100 because in some instances an application was assigned multiple withdrawal codes.
[a]See appendix III for a description of the codes included in the "All others" category.

Figure 10. Frequency of SBA Withdrawal Codes for Hurricane Sandy Business Disaster Loan Applications by Loan Type.

The three other most prevalent reasons for withdrawals were an applicant's request to withdraw due to a change of plans (852 withdrawals, or 18 percent); SBA's withdrawal due to IRS having no record (582 withdrawals, or 12 percent); and an applicant's request to withdraw due to other reasons (553 withdrawals, or 12 percent). See figure 10 for more detail. Of the 4,180 business loan applications SBA approved for Hurricane Sandy, 1,578 (38 percent) were cancelled as of January 31, 2014—a cancellation rate higher than those of the previous disasters we examined.[42] The cancellation rates for these previous disasters ranged from approximately 22 percent (for Hurricane Wilma), to approximately 30 percent (for Hurricane Ike and Katrina), as shown in figure 11.

Of the 1,578 business loans cancelled after Hurricane Sandy, 1,171 loans (74 percent) were cancelled at the borrower's request, while 407 (26 percent) were cancelled by SBA. The most prevalent cancellation code in either category was "Other reasons – (borrower's decision)," which accounted for 720 (46 percent) of all cancellations.[43] The second most prevalent code—and the leading code in the category of SBA-initiated cancellations—was "Failure to complete and return all loan closing documents," representing 336 cancellations (21 percent). See figure 12 for more detail.

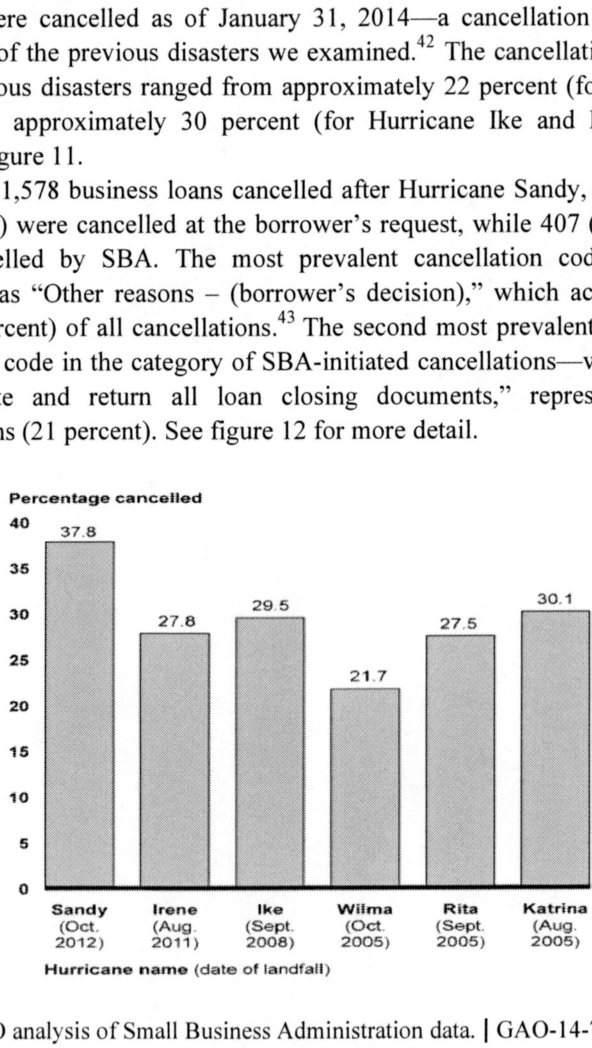

Source: GAO analysis of Small Business Administration data. | GAO-14-760.

Figure 11. Comparison of Business Loan Cancellation Rates for Hurricane Sandy Compared with Previous Selected Disasters, 2005-2013.

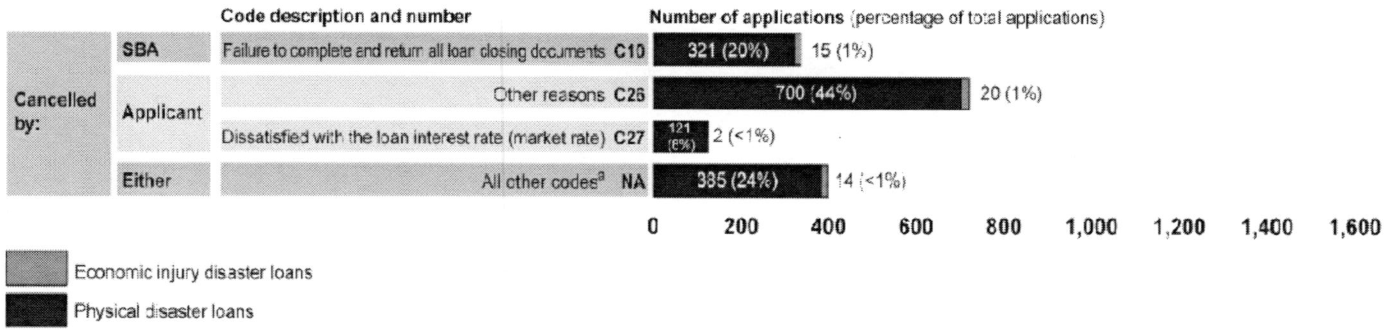

Source: GAO analysis of Small Business Administration data. | GAO-14-760.

Figure 12. Frequency of SBA Cancellation Codes for Hurricane Sandy Business Disaster Loans by Loan Type.

Half of the entities we spoke with—selected SBDCs and local business organizations such as chambers of commerce—in New York and New Jersey provided perspectives on the most common reasons why applications were withdrawn following Hurricane Sandy.[44] Some entities we spoke with noted that a common reason for application withdrawal was a change of plans with regard to how business owners were going to fund their recovery. For example, some entities said that businesses withdrew their SBA loan applications because alternate sources of funding were made available, including insurance claim proceeds and state grant programs. Entities also noted other reasons for application withdrawal, such as frustration with waiting for loan applications to be processed and a desire to not incur additional debt.

According to SBA, factors affecting the withdrawal and cancellation rates for Hurricane Sandy include the footprint of the disaster area and the availability of alternative sources of recovery (e.g., insurance coverage and grants). The Hurricane Sandy footprint covered areas with higher rates of insurance coverage compared to previous disasters. Additionally, officials told us that the rollout of programs funded by the Department of Housing and Urban Development's Community Development Block Grant program (CDBG) began earlier than in past disasters, and that state grantees—specifically New Jersey and New York—obtained CDBG funds and accepted applications for their respective state grant programs shortly after the disaster struck.

SBA Has Not Implemented Three Guaranteed Disaster Loan Programs Required by Law

More than 6 years after Congress passed the Small Business Disaster Response and Loan Improvements Act of 2008, SBA has not piloted nor implemented three guaranteed disaster loan programs, which were therefore unavailable in response to Hurricane Sandy. Therefore, the potential effectiveness of these programs remains unclear. As previously discussed, the act mandated the creation of the Immediate Disaster Assistance Program (IDAP), the Expedited Disaster Assistance Loan Program (EDALP), and the Private Disaster Assistance Program (PDAP). According to SBA officials, the agency opted to implement IDAP first, because the loan limit is lower than that of EDALP and PDAP and funds were appropriated to pilot this program.

A majority of entities with whom we spoke said that a program like IDAP (i.e., a loan up to $25,000 with a 36-hour application approval time) would be useful, as it could provide business owners with a short-term infusion of capital to complete tasks such as debris removal, repairing physical damage, and purchasing inventory. As previously discussed, on average, it took SBA 45 and 38 days to process physical business disaster loan and EIDL applications resulting from Hurricane Sandy, respectively (from application receipt to loan decision). As such, it took additional time for funds to be disbursed to an applicant approved for a direct disaster loan—time that could impact whether a business seeks other financing options or remain in operation.

In a July 2009 report, GAO reported that SBA was planning to implement pilot programs for IDAP and EDALP to test applicable program requirements and to see how private lenders would administer the programs.[45] GAO also reported that SBA requested funding to carry out the requirements for these two programs in the President's budget for fiscal year 2010. SBA received subsidy and administrative cost funding totaling $3 million in the 2010 appropriation, which would allow the agency to pilot test about 600 loans under IDAP[46] The agency also issued regulations for IDAP in October 2010. In May 2010, SBA also told GAO that its goal was to have the pilot for IDAP in place by September 2010.[47] In addition, a SBA report noted in June 2012 that IDAP would be ready to be put into operation by the end of that year. However, as of August 2014, the pilot program for IDAP had not been conducted.

According to SBA officials, the program has not been implemented for two primary reasons: (1) information technology challenges and (2) feedback from lenders indicating that program requirements may hinder lender participation in the program. First, the electronic systems that would be used to process IDAP applications did not interface smoothly with one another. According to SBA officials, IDAP's readiness was, in part based on the ability of E-Tran (the 7(a) program's electronic loan processing system) to interface with DCMS (the Disaster Loan Program's electronic loan processing system).[48] Officials said that a new information technology system is being developed—SBA One—and that it is more efficient to make DCMS interoperable with the new system to process IDAP applications than to make enhancements to E-Tran.[49] SBA anticipates that SBA One will be operational by early 2015.

Second, SBA told us that they received feedback from lenders indicating challenges that may discourage lenders from participating in the program, although SBA's documentation of this feedback is limited. In March 2010, SBA organized a forum with 11 lenders in the Gulf Coast to obtain their views on IDAP. Lenders stated that the program had to have simple eligibility determination, as well as confirmation that a potential IDAP borrower had applied for an SBA disaster loan before the lender would approve an IDAP loan.[50] Lenders also expressed concerns about the possibility of guarantee denials if an applicant did not take out an SBA disaster loan.[51] In addition, according to SBA, in 2010 the agency also consulted with Iowa lenders in flood-prone regions within the state via conference calls and lenders expressed similar concerns about IDAP. However, both the Gulf Coast forum and the conference calls with Iowa lenders were not documented at the time the lender outreach was conducted. As such, SBA officials must rely on the memory of staff who were present for the discussions. In response to our request for information on SBA's outreach to these lenders, in July 2014, SBA provided a one-page summary on the outreach to both the Gulf Coast and Iowa lenders. The summary included a list of the Gulf Coast lenders that participated but not the Iowa lenders, and the discussion of lenders' concerns was limited.

In addition, according to SBA officials, in November 2012 the agency solicited informal feedback from lenders in the Hurricane Sandy-affected area about the usefulness of IDAP and its features. According to SBA officials, lenders were concerned about the statutory requirement that provides an applicant a minimum of 10 years to repay the IDAP loan if a loan through the Disaster Loan Program is not approved. 52 Lenders also expressed disinterest in servicing a small loan amount (up to $25,000) for a term that long. SBA officials noted that small-dollar loans such as those made under IDAP are not products that lenders typically offer. In addition, SBA's IDAP regulations allow a lender to charge a borrower an optional application fee to recoup some of the lender's loan processing costs, but the one-time fee may not exceed $250 and an IDAP lender generally may not charge a borrower any additional fees.[53] According to SBA officials, they also did not document lender feedback from this outreach effort. SBA officials told us that feedback on IDAP requirements was obtained from three banks, although officials could recall the identity of only one bank.

In July 2014, SBA officials told us that the agency is still trying to conduct the IDAP pilot by attempting to identify solutions to increase lender participation. However, officials also noted that the lenders they met with were not willing to participate in IDAP (or an IDAP pilot) without changes to the statutory servicing term and the SBA regulatory program fee. Based on lender feedback, SBA officials said that the current statutory requirements, such as the 10-year loan term, make a product like IDAP undesirable and that lenders are not likely to participate in IDAP unless the loan term is decreased—for example, to between 5 and 7 years. However, revising statutory program requirements requires congressional action. SBA officials said that they have not discussed the feedback received from lenders with Congress. As a result, Congress does not have this information about SBA's challenges with and plans to pilot and implement IDAP. Further, SBA officials told us that the agency plans to use IDAP as a guide to develop EDALP and PDAP, and until challenges with IDAP are resolved, SBA does not plan to implement these two programs.

In addition, federal internal control standards state that all transactions and other significant events should be promptly recorded to maintain their relevance and value to management in controlling operations and making decisions. Documentation should also be readily available for examination.[54] Although SBA obtained feedback from lenders on their willingness to participate in IDAP, documentation of these discussions was limited and SBA has not conducted a formal documented evaluation of lenders' feedback that would establish the basis for proposed changes to requirements for Congress to consider. Without an appropriately documented evaluation of its outreach to lenders, SBA may not have complete and reliable information on which to base its reporting to Congress about its challenges with implementing the programs required by the act.

More generally, we previously recommended that SBA develop an implementation plan and report to Congress on the agency's progress in addressing all requirements within the act and include milestone dates for completing implementation and any major program, resource, or other challenges the agency faces as it continues efforts to address requirements in the act.[55] As of September 2014, the agency had not implemented this recommendation. Not having an implementation plan in place for addressing the remaining requirements and not reporting to Congress on related challenges can lead to a lack of transparency about the agency's progress toward implementing these programs and its ability to adequately prepare for and respond to disasters.

CONCLUSION

Following Hurricane Sandy, SBA did not meet its application processing goal for a number of reasons. SBA said it was challenged by a high volume of loan applications early in its response to the disaster primarily due to the use of electronic applications, as well as technological and other issues. Although electronic loan applications were created to simplify and expedite the application process and SBA has continued to encourage submitting applications electronically, SBA did not anticipate receiving such a large volume of electronic loan applications early in its response following Hurricane Sandy and was unprepared to process them. While SBA officials told us the agency has taken steps to respond to these challenges, it has not revised its disaster planning documents— including the Disaster Preparedness and Recovery Plan and *ODA Playbook*—to reflect the unexpected volume of early applications and its potential impact on staffing, resources, and forecasting models. Federal internal control standards state that management should identify risks and take action to manage them. Without taking the large volume of applications it received shortly after Hurricane Sandy into account in its disaster planning documents and analyzing the risk that trend may pose for timely disaster response, SBA risks being unprepared for a similar experience following future disasters, which may result in delays in providing loan funds to disaster victims.

Due to lender feedback and technology challenges, SBA has yet to pilot IDAP and has not implemented IDAP, EDALP, and PDAP as required by Congress in the Small Business Disaster Response and Loan Improvements Act of 2008. SBA has also not developed an implementation plan for addressing the act's requirements, as GAO recommended in 2009. This plan was to include, among other things, challenges the agency faces in implementing the act's requirements. Private sector lenders have given SBA feedback on barriers to their willingness to participate in IDAP, such as statutory program requirements that could make the make the loans undesirable for lenders. However SBA has not conducted a formal documented evaluation of lenders' feedback that would establish the basis for proposed changes to requirements for Congress to consider. In addition, SBA has not provided lenders' feedback and its evaluation of that feedback to Congress—the entity that could remove potential barriers to SBA's implementation of the program. As a result, SBA may lack reliable information to share with Congress about what program requirements should be revised to encourage lender participation. Such information could, for example, be obtained by conducting further outreach to lenders and documenting this outreach in accordance with federal internal control standards. Without sharing information with Congress on challenges to implementing IDAP, SBA may continue to face difficulties in implementing programs intended to provide assistance to disaster victims.

RECOMMENDATIONS FOR EXECUTIVE ACTION

In order to address potential risk to SBA's ability to provide timely disaster assistance in the future, based on the agency's experience from Hurricane Sandy, we recommend that the Administrator of SBA direct the Office of Disaster Assistance to take the following action:

- Revise SBA's disaster planning documents to anticipate the potential impact of early application submissions on staffing and resources for future disasters, as well as the risk this impact may pose for SBA's timely disaster response.

In order to provide Congress with reliable information on challenges SBA has faced in implementing IDAP, we recommend that the Administrator of SBA direct the Office of Capital Access to take the following two actions:

- Conduct a formal documented evaluation of lenders' feedback that can inform SBA and Congress about statutory changes that may be necessary to encourage lenders' participation in IDAP.
- Report to Congress on the challenges SBA has faced in implementing IDAP and on statutory changes that may be necessary to facilitate SBA's implementation of the program.

AGENCY COMMENTS

We provided a draft of this report to SBA for review and comment. SBA provided written comments, which are described below. SBA generally agreed with all recommendations presented in this report. In response to SBA's comments addressing our first recommendation, we added information to the report on the role of electronic application submissions in the unanticipated large volume of applications SBA received early in its response to Hurricane Sandy.

In response to our recommendation that SBA revise its disaster planning documents to anticipate the potential impact of early application submissions on staffing and resources for future disasters, SBA agreed with the importance of incorporating this information into its planning documents. In addition, SBA stated that it has already taken steps to document and address the early volume of application submissions. For example, SBA noted that the most recent Disaster Planning and Recovery Plan (DPRP) released in June 2014 addresses the increased usage of the electronic application and the resulting impact of early application submissions on staffing and resources due to both the convenience and speed of the Internet-based application and to the elimination of postal handling time. SBA further stated that it believes the agency has already met this recommendation. While SBA has acknowledged the impact of early application submissions on staffing and resources, SBA has not yet developed details of how it will incorporate its experience with Hurricane Sandy throughout its disaster planning to better address resource and staffing needs resulting from early application submissions. Thus, we maintain that SBA should take additional action to fully address this recommendation. SBA stated that it plans to go further and incorporate the early application submissions in the next update of the *ODA Playbook* and in other ODA planning documents. It will be important for SBA to follow through on these plans in order to reduce the risk that increased early application submissions pose to SBA's ability to deliver timely disaster assistance.

In response to our recommendations that SBA conduct a formal documented evaluation of lenders' feedback on IDAP, and report to Congress on challenges to implementing the program, SBA agreed that its Office of Capital Access (OCA) has not performed a formal documented evaluation of IDAP. SBA further stated that OCA intends to conduct such an evaluation that will include feedback from a variety of lenders and will use the information to identify potential changes to the statute, current regulations, procedures, and forms for IDAP.

SBA acknowledged that the focus of our report was on its assistance to small businesses following Hurricane Sandy; however, SBA noted that its mission to provide disaster loan assistance is not limited to small businesses but also extends to businesses of all sizes, private nonprofit organizations, homeowners, and renters. Specifically, SBA noted that about 83 percent of loan applications received after Hurricane Sandy were for home disaster loans, which caused a significant impact on the agency's overall response. In our report, which specifically reviewed SBA's assistance to small businesses following Hurricane Sandy, we acknowledge that SBA disaster assistance is not limited to small businesses and home loan applications comprised about 80 percent of all Hurricane Sandy disaster applications. Our report also notes that business owners could have faced delays due to the large number of home loan applications submitted ahead of them and that one process improvement SBA made after Hurricane Sandy was to revise its loan processing approach to create two separate tracks for home and business disaster loan applications.

Finally, consistent with information provided in the report, SBA summarized various other process improvements that the agency has implemented to mitigate the challenges faced during Hurricane Sandy and for future disasters. The draft report that we sent to SBA for comment also described many of these changes. For example, we note in our report that SBA is taking several steps to streamline the documentation requirements for applicants and improve the process to submit additional information. However, as stated in our report, because SBA has not received a large volume of loan applications since Hurricane Sandy, it is too early to determine the extent to which these changes will improve the timeliness of SBA's response to future disasters with a magnitude similar to that of Hurricane Sandy.

Sincerely Yours,
William B. Shear
Director,
Financial Markets and Community Investment

APPENDIX I: OBJECTIVES, SCOPE, AND METHODOLOGY

Our objectives were to examine (1) the timeliness of the Small Business Administration's (SBA) disaster assistance to small businesses following Hurricane Sandy and the factors that affected SBA's timeliness, (2) the loan approval rates for small businesses and reasons for decline following Hurricane Sandy, in comparison with those of previous disasters, and (3) the extent to which SBA has implemented loan programs mandated by the Small Business Disaster Response and Loan Improvements Act of 2008.

To evaluate the timeliness of disaster assistance to small businesses, we obtained and analyzed data from SBA's Disaster Credit Management System (DCMS) on the average number of days it took SBA to process Hurricane Sandy business loans. Specifically, we analyzed the number of days from application receipt to loan decision, loan approval to closing, and closing to first loan disbursement.[1] For analyses on the timeliness of disaster assistance, we used information on all loan application decisions, including information on reconsidered and appealed applications.[2] To assess whether SBA met its timeliness goal of processing applications, we reviewed key documents, such as SBA's Disaster Preparedness and Recovery Plan (DPRP), to identify any timeliness goals SBA has established and compared the DCMS data with such goals. Further, we reviewed work conducted by the SBA Office of Inspector General (OIG) to assess any changes SBA has made to setting and communicating goals related to timely disaster-loan processing.

To identify factors that affected SBA's timeliness of disaster assistance following Hurricane Sandy, we reviewed an internal assessment of SBA's response to Hurricane Sandy as well as key SBA documents that outline agency roles and responsibilities for responding to a disaster, including the Disaster Assistance Program Standard Operating Procedures, DPRP, and Office of Disaster Assistance (ODA) *Playbook*. We also obtained and analyzed DCMS data on the number of business loan applications received after Hurricane Sandy by type of submission method (electronic or paper) and in processing from October 2012 to January 2014 to assess the extent to which identified factors affected SBA's timeliness. Further, we interviewed SBA officials from the ODA, Processing and Disbursement Center, and Office of Disaster Planning to obtain their perspectives on challenges that SBA faced in providing disaster assistance and efforts taken to respond to those challenges.

To obtain information on the factors that affected the timeliness of SBA's assistance from the perspective of small businesses applying for loans, we spoke with selected Small Business Development Centers (SBDC) and other local organizations, such as chambers of commerce in New York and New Jersey, the states most heavily impacted by Hurricane Sandy.[3] These selected entities were located in the counties that (1) were highly impacted based on assessments and reports by the Federal Emergency Management Agency (FEMA) and Department of Housing and Urban Development and (2) generated the most SBA business loan applications after Hurricane Sandy.[4] We attempted to contact at least one SBDC and one local organization in each selected county. However, one SBDC and the local business organizations contacted in one county declined to be interviewed. Therefore, we spoke to a total of 14 entities (6 SBDCs and 8 local business organizations). While the results of these interviews could not be generalized to all counties and states affected by Hurricane Sandy, they provided insight into challenges SBA faced in providing timely assistance to small businesses.

To calculate the loan approval rates for small businesses following Hurricane Sandy, we obtained and analyzed DCMS data on approvals and declines. Specifically, we divided the number of approved applications by the total number of approved and declined applications.[5] Because physical disaster loans are not restricted to small businesses, we were not able to determine the physical disaster loan approval rate for small businesses.

To evaluate the reasons why business loan applications were declined or withdrawn following Hurricane Sandy, we obtained DCMS data on the codes applied to declined and withdrawn applications and identified the most prevalent codes. To identify the reasons why approved business loans were cancelled following Hurricane Sandy, we obtained DCMS data on the codes applied to cancelled loans and identified the most prevalent cancellation codes.

To compare the application approval and withdrawal rates, the reasons for decline, and the loan cancellation rates following Hurricane Sandy with those of previous disasters, we selected the five largest disasters since 2005 based on disaster loan application volume and performed the same analyses of the relevant DCMS data. The five disasters we selected were Hurricanes Katrina, Rita, Wilma, Ike, and Irene. We interviewed SBA officials to confirm that the Disaster Loan Program's procedures for how data elements are defined, entered, and maintained remained substantially unchanged during this time period.

To assess the reliability of DCMS data, we (1) performed comparisons across the data SBA provided to identify any obvious errors in accuracy and completeness; (2) reviewed related documentation such as the DCMS data dictionary and data verification documentation; and (3) worked with appropriate agency officials to identify any data problems. When we found discrepancies, such as unpopulated fields or data entry errors, we notified agency officials and worked with these officials to correct the discrepancies before conducting our analysis. We determined that the data were sufficiently reliable for the purposes of our report.

To examine the extent to which SBA has implemented loan programs mandated by the Small Business Disaster Response and Loan Improvements Act of 2008, we reviewed the act, other relevant legislation, and draft guidance for the Immediate Disaster Assistance Program. Further, we interviewed SBA officials from ODA and the Office of Capital Access to obtain their perspectives on SBA's efforts to implement the Immediate Disaster Assistance Program (IDAP) and related challenges because IDAP is the only one of the three mandated guaranteed loan programs for which SBA has issued regulations.

We conducted this performance audit from August 2013 through September 2014 in accordance with generally accepted government auditing standards. Those standards require that we plan and perform the audit to obtain sufficient, appropriate evidence to provide a reasonable basis for our findings and conclusions based on our audit objectives. We believe that the evidence obtained provides a reasonable basis for our findings and conclusions based on our audit objectives.

APPENDIX II: NEW JERSEY AND NEW YORK HURRICANE SANDY BUSINESS DISASTER LOAN APPLICATIONS RECEIVED AND APPROVED AND DOLLAR AMOUNTS (AS OF JANUARY 2014)

The table below outlines the number of Hurricane Sandy business disaster loans that were received and approved and the dollar amounts of approved loans for highly impacted counties in New York and New Jersey. This table does not include all counties that were impacted by Hurricane Sandy and all counties that received SBA business disaster loans.

Table 4. Hurricane Sandy Business Disaster Loan Applications Received and Approved and Dollar Amounts for Highly Impacted Counties in New Jersey and New York (as of January 2014)

State	County	FEMA impact category[a]	Total SBA business applications received	Total SBA business applications approved[b]	Dollar amount of approved loans
New Jersey	Ocean	Very high	2,217	732	$68,073,800
New York	Nassau	Very high	2,113	588	$61,197,400
New York	Kings	Very high	1,848	458	$71,496,600
New Jersey	Monmouth	Very high	1,252	370	$36,062,400
New York	Queens	Very high	1,523	370	$56,342,600
New York	Suffolk	Very high	859	237	$26,445,100
New York	New York	Very high	1,251	236	$30,745,300
New York	Richmond	Very high	720	192	$22,518,900
New Jersey	Hudson	Very high	678	175	$24,307,800

State	County	FEMA impact category[a]	Total SBA business applications received	Total SBA business applications approved[b]	Dollar amount of approved loans
New Jersey	Atlantic	Very high	653	136	$16,170,400
New Jersey	Cape May	Very high	346	122	$8,857,800
New Jersey	Bergen	Very high	384	104	$17,505,600
New Jersey	Middlesex	Very high	278	80	$11,232,500
New Jersey	Essex	Very high	266	50	$6,215,200
New Jersey	Union	Very high	189	39	$6,263,400
New York	Bronx	Very high	111	31	$3,786,000
New York	Westchester	High	112	28	$2,864,000
New York	Rockland	High	78	20	$1,908,600
New Jersey	Somerset	High	50	15	$949,100
New Jersey	Passaic	High	77	14	$1,663,500
New Jersey	Morris	High	75	13	$1,343,100
New Jersey	Mercer	High	33	9	$686,700
New Jersey	Cumberland	High	25	5	$1,029,200
New Jersey	Burlington	High	19	4	$263,200
New Jersey	Camden	High	8	0	$0
New Jersey	Gloucester	High	1	0	$0
New Jersey	Salem	High	4	0	$0

Source: GAO analysis of SBA data. | GAO-14-760

[a]The Federal Emergency Management Agency (FEMA) defines "Very high" impact areas as those in which more than 10,000 of the county's population were exposed to the storm's surge. "High" impact areas are those in which 500 to 10,000 of the county's population were exposed to the storm's surge, modeled wind damage was greater than $100 million, or precipitation was greater than 8 inches.

[b]In addition to applications that were approved and declined, "Total applications received" includes applications that were ultimately withdrawn, which are not factored into SBA's calculation of approval rates.

APPENDIX III: EXPLANATION OF DISASTER LOAN PROGRAM'S DECLINE CODES, WITHDRAWAL CODES, AND CANCELLATION CODES

The following tables include decline, withdrawal, and cancellation codes—as described in SBA's Standard Operating Procedures, Appendixes 3, 4, and 9—that are used for physical disaster loan applications (both home and business), and Economic Injury Disaster Loan applications.

Table 5. Decline Codes

Decline Code 20	Lack of repayment ability – Applicant's income below minimum income level for the family size • (Used in Summary Decline, Auto-Decline, and Pre-LV Review processes only.)
Decline Code 21	Lack of repayment ability
Decline Code 22	Lack of ability to repay a disaster loan within a maximum seven-year term • (Used only for business physical loans with credit available elsewhere. Does not apply to nonprofits.)
Decline Code 23	Inadequate cash flow to repay a disaster loan and meet other obligations
Decline Code 24	Excessive amount of debt relative to net worth • (Never used as only reason for decline)
Decline Code 25	Inadequate working capital even if SBA could approve a loan • (Never used as only reason for decline)
Decline Code 26	Unsatisfactory history on an existing or previous SBA loan
Decline Code 27	Unsatisfactory history on a Federal obligation
Decline Code 28	Unsatisfactory credit history
Decline Code 29	Unsatisfactory debt payment history • (Used for other than a credit bureau)
Decline Code 30	No disaster-related damage • (Used only when the verified loss is zero)
Decline Code 31	Economic injury is not substantiated
31-A	No Needs
31-B	Disaster Gross Margin Exceeds Normal
31-C	Custom Text
Decline Code 32	Business activity is not eligible • (For EIDLs only)
Decline Code 33	Not eligible because the applicant is not a small business • (For EIDLs only)
Decline Code 34	Credit is available elsewhere • (For EIDLs only)
Decline Code 35	Not located in the declared disaster area
35-A	For physical applications
35-B	For EIDL applications
Decline Code 36	Ineligible real property • (To be used for secondary homes, etc.)
Decline Code 37	Ineligible personal property
Decline Code 38	Not eligible due to recoveries from other sources
Decline Code 39-A	Not eligible due to failure to maintain flood insurance coverage on an existing SBA loan
Decline Code 39-B	Not eligible due to failure to maintain required flood insurance on a loan from a federally regulated lender
Decline Code 39-C	Not eligible due to failure to maintain required flood insurance as directed by the Federal Emergency Management Agency (FEMA)

Decline Code 40	Not a qualified business
40-A	Business
40-B	Rental
Decline Code 41	Refusal to pledge available collateral
Decline Code 42	Not eligible due to delinquent child support payments
Decline Code 43	Not eligible due to character reasons
Decline Code 44-I	Lack of repayment ability – Below minimum income level for the family size based upon the applicant's income alone
Decline Code 44-R	Lack of ability to repay a disaster loan based upon the applicant's income alone
Decline Code 45	Not eligible due to an outstanding judgment lien for a Federal debt
Decline Code 46-A	Agricultural enterprises are not eligible
Decline Code 46-B	Members of a fishing crew do not qualify as an eligible small business concern
Decline Code 46-C	Not eligible due to property being located in a Coastal Barrier Resource Area
Decline Code 46-D	Custom text

Source: SBA | GAO-14-760

Table 6. Withdrawal Codes

Withdrawal Code 51	Requested information was not furnished
Withdrawal Code 52	Applicant's Request – A change in plans
Withdrawal Code 53	Applicant's Request – No reason given
Withdrawal Code 54	Applicant's Request – Due to availability of insurance or other recovery
Withdrawal Code 55	Applicant's Request – State basis for request[a]
Withdrawal Code 56-A	Unable to verify property
Withdrawal Code 56-B	Custom text
Withdrawal Code 57	Consolidation of multiple applications
Withdrawal Code 58	Consolidation of related applications
Withdrawal Code 59	IRS has no record
Withdrawal Code 60	Character eligibility determination
• 60-a	• Withdrawal of an otherwise approvable application
• 60-d	• Decline[b]
• 60-w	• Withdrawal[c]
Withdrawal Code 61	Applicant's Request – Due to market rate

Source: SBA | GAO-14-760

[a]According to an SBA official, the most prevalent examples of Withdrawal Code 55 reasons are: "Borrower does not want the loan" and "Borrower needs more time to gather additional information being requested."

[b]Although this code is listed among the withdrawal codes, SBA applies this code to declined applications, rather than withdrawals.

[c]Unlike code 60-a, code 60-w is applied when SBA cites independent reason for withdrawing the application. Code 60-a is reserved for cases in which a character determination is the sole reason for withdrawal.

Table 7. Cancellation Codes

Agency Cancellation	
C10	Failure to complete and return all loan closing documents
C11	Failure to satisfy all terms and conditions of the loan
C12	Adverse change - IHP referral
C13	Adverse change - Other
C14	Subsequent recoveries exceed verified loss
C15	Did not need all the funds – (Agency Decision)
C16	Other reasons – (Agency Decision)
Cancellation at Borrower's Request	
C20	Adequate recovery from other sources
C21	Reluctant to incur additional debt
C22	Dissatisfied with loan terms and conditions
C23	Dissatisfied with insurance requirements
C25	Did not need all the funds – (Borrower Decision)
C24	Unwilling to pledge collateral
C26	Other reasons – (Borrower Decision)a
C27	Dissatisfied with loan interest rate (market rate) – (Borrower Decision)

Source: SBA. | GAO-14-760

[a] According to an SBA official, the most prevalent examples of Cancellation Code 26—Other reasons are: "Borrower does not want the loan", "No reason given," and "Borrower is not ready to do the project."

End Notes

[1] Pub. L. 113-2, Division A, 127 Stat. 27, 39 (2013).

[2] Pub. L. No. 110-246, title XII, subtitle B, 122 Stat. 2168 (2008).

[3] In addition to the President, SBA can also declare an area to be a disaster zone. As a result of these "administratively declared disasters," residents and businesses in the affected areas are eligible for SBA disaster loans, but generally not for other forms of federal assistance.

[4] SBA's income thresholds vary by household size and are updated on an annual basis.

[5] Response and Recovery are two of the five "national preparedness frameworks" set forth in the federal government's comprehensive approach to preparedness and domestic disaster incident management. The other three frameworks are Prevention, Protection, and Mitigation. The Response framework refers to capabilities necessary to save lives, protect property and the environment, and meet basic human needs after an incident has occurred. The Recovery framework refers to capabilities necessary to assist communities affected by an incident to recover effectively, including, but not limited to, rebuilding infrastructure systems; providing adequate interim and long-term housing for survivors; restoring health, social, and community services; promoting economic development; and restoring natural and cultural resources.

[6] An agricultural enterprise is a business primarily engaged in the production of food and fiber, ranching and raising livestock, aquaculture, and all other farming and agriculture-related

industries. 13 C.F.R. § 123.201(a). Additional eligibility restrictions can be found at 13 C.F.R. §§ 123.101, 123.201. For more information, see GAO, *Disaster Assistance: USDA and SBA Could Do More to Help Aquaculture and Nursery Producers,* GAO-12-844 (Washington D.C.: Sept. 11, 2012).

[7] SBA's size standards define whether a business is "small." Size standards have been established for types of economic activity, or industry, generally under the North American Industry Classification System. 13 C.F.R. Part 121. "Credit elsewhere" refers to circumstances in which SBA believes a business has the availability of credit from nonfederal sources on reasonable terms and conditions because of its cash flow and disposable assets. 13 C.F.R. §§ 123.104, 123.300.

[8] Agricultural enterprises are also not eligible for economic injury disaster loans. Certain small nurseries affected by a drought designated by the Secretary of Agriculture, small agricultural cooperatives, and producer cooperatives are eligible for economic injury disaster loans. 13 C.F.R. § 123.300(c).

[9] Businesses of all sizes are eligible for physical disaster loans, and SBA does not maintain data on business size for these types of loans. Therefore, information specifically for small businesses that received a physical disaster loan is unavailable. Data on physical disaster loan applications or total business disaster loan applications (physical loan applications and EIDLs) presented in this report include information about businesses of all sizes.

[10] For the purposes of this report, the term "physical disaster loan" refers to any loan resulting from a physical disaster loan application, which may or may not include loan funds for economic injury, while the term "EIDL" refers only to those loans that result from EIDL applications. Unless otherwise noted, our analyses are restricted to businesses.

[11] To fulfill SBA documentation requirements, each principal owning 20 percent or more of the applicant business and each general partner or managing member must also provide a signed personal financial statement and Form 8821. Further, each affiliate of the business must provide a signed Form 8821. Affiliates include, but are not limited to, business parents, subsidiaries, or other businesses with common ownership or management. For the complete list of required documents for a business disaster loan application, see SBA's website, http://www.sba.gov/content/disaster-loan-paper-applications, accessed on September 4, 2014.

[12] SBA publishes notices of disaster declarations in the *Federal Register*, which generally includes the application deadline and the location for filing a loan application. 13 C.F.R. § 123.3(b).

[13] 13 C.F.R. §§ 123.6. This regulation was revised in April 2014 to allow SBA to place greater emphasis on an applicant's credit score in determining repayment ability.

[14] 13 C.F.R. §§ 123.201(a), 123.301.

[15] According to SBA's procedures, an acceptable application is one that has a signed and reasonably completed application form and a fully completed and signed Tax Information Authorization (IRS Form 8821) for each required taxpayer or entity. SBA returns unacceptable applications and requests the information needed to make the application acceptable.

[16] Application processing functions from application entry, scanning, and processing through to a decision occur in the PDC, which is located in Ft. Worth, Texas. All loan approvals, loan document generation, loan closing, and disbursement of loan proceeds are functions of the PDC.

[17] Automatic declines occur when an applicant's credit indicates that a credit decline decision is obvious, and DCMS automatically declines these applications. Pre-loss verification declines

occur when SBA identifies applications having a very low likelihood of being approved based on the applicant's credit, repayment ability, or other eligibility restrictions.

[18] Duplication of benefits refers to a general prohibition on receiving disaster assistance for a loss if the loss has already been compensated from another source, such as insurance. See 42 U.S.C. § 5155.

[19] SBA officials stated that to determine repayment ability for a business, SBA looks at several pieces of information, including tax returns, net income, cash flows, and assets, and calculates if the business has enough cash to service additional debt.

[20] Secured loans are those which require collateral.

[21] The total number of business disaster loan applications includes "reconsiderations"—applications declined upon first submission and resubmitted for a second review—and "appeals"—reconsidered applications that are declined and resubmitted again for a third review.

[22] Pub. L. No.110-246, §§ 12083 -12085.

[23] According to SBA officials, this is an internal timeliness goal. Per the DPRP, SBA is positioned to handle up to 100,000 disaster loan applications within its performance goal of 21 days (from receipt of the application to decision). According to SBA officials, it becomes increasingly challenging to meet this 21-day goal once disaster loan application volume exceeds about 50,000 applications. Hurricane Sandy generated 88,381 total disaster loan applications.

[24] In a June 2014 report, the SBA Office of Inspector General (OIG) found that the methodology that SBA used to calculate average loan processing times incorporated the processing times for automatically declined applications. Further, the OIG found that SBA's disaster loan processing time goals—particularly those reported in the agency's Congressional Budget Justification—were established without considering potential increases in application volumes. The OIG recommended, among other things, that SBA establish processing time standards for different application volumes based on historical performance. SBA agreed to establish such processing time standards and has included these standards in the most recent update of ODA's *Playbook* and DPRP. See SBA, Office of Inspector General, *Improving Accuracy of Performance Reporting to Better Manage Disaster Loan Processing Time Expectations*, Report No. 14-14 (Washington, D.C.: June 30, 2014).

[25] SBA extended the deadlines for five states. Therefore, though Hurricane Sandy made landfall in October 2012, several states' deadlines were extended to July-September 2013.

[26] The DPRP is an agency-wide disaster response plan whereas the *ODA Playbook* is an internal document to outline the roles and responsibilities of ODA departments, resource partners, and other partners in the private sector at each major phase of the disaster recovery process.

[27] GAO, *Standards for Internal Control in the Federal Government*, GAO/AIMD-00-21.3.1 (Washington, D.C.: November 1999).

[28] While the PDC is a permanent loan application processing center, SBA maintains space for additional loan officers in its Sacramento office, known as the Disaster Assistance Field Operations Center—West. Following Hurricane Sandy, ODA also added loan officers in the Disaster Assistance Customer Service Center in Buffalo, which typically provides customer support services such as conducting outreach calls and providing customers with information.

[29] According to the DPRP, an increase in staff is typically initiated once ODA decides whether a disaster can be handled by the agency's internal capacity. However, an increase in staff is not implemented merely because a specified number of applications has been projected.

Increasing staff is a specific decision to be made by the Administrator after considering the overall nature of the disaster.

[30] According to SBA, homeowners normally apply for SBA disaster loans faster than business owners. Business owners first assess the economic damage to their businesses caused by disrupted supply chains, displaced consumers, and other factors.

[31] The Hurricane Sandy Rebuilding Task Force was created in December 2012 by President Obama to ensure cabinet-level, government-wide and region-wide coordination to help communities affected by Hurricane Sandy as they are making decisions about long-term rebuilding. The Task Force was chaired by the U.S. Department of Housing and Urban Development, and SBA participated as a member. In August 2013, the Task Force released the Hurricane Sandy Rebuilding Strategy report, which included 69 recommendations across several policy areas that are designed to eliminate barriers to recover while ensuring effectiveness and accountability, among other things. See Hurricane Sandy Rebuilding Task Force, *Hurricane Sandy Rebuilding Strategy: Stronger Communities, A Resilient Region* (Washington, D.C.: August 2013).

[32] Prior to its amendment, 13 C.F.R. § 123.6 required SBA to analyze every applicant's personal and business cash flow, a time-consuming process that includes debt reconciliation and a repayment analysis to determine if there are funds available for both loan payments and day-to-day living expenses. SBA has amended 13 C.F.R. § 123.6 to allow it to base its repayment ability determination on either the applicant's cash flow or credit, including credit score. The repayment analysis will still include the verification of income/employment through federal income tax returns. Additionally, SBA still plans to analyze personal or business cash flow to determine repayment ability for those applicants that do not have strong credit, as determined by SBA.

[33] SAPP was developed after ODA headquarters staff recognized that an above-average percentage of applications with higher household income levels and credit scores were being submitted following Hurricane Sandy. ODA received approval to pilot a program that would expedite processing for these applications 1 month into the Hurricane Sandy response for home loan applications only.

[34] According to SBA, a loan officer is responsible for making loan application decisions and a case manager is responsible for closing and disbursing the loan.

[35] Business loan applications consist of both physical loan applications and EIDL applications. Physical loan applications can also be considered for economic injury loan proceeds. The loans that result from these applications can therefore involve both a physical loan component and an EIDL component.

[36] Approval rates are calculated by dividing the number of approved applications by the sum of approved and declined applications. Applications that are withdrawn—when a loan application is removed from consideration before SBA has rendered an approval decision—are not included in this calculation, and therefore do not impact the approval rate. We are unable to report on SBA's approval rates for small business applicants for physical disaster loans because these loans are not limited to small businesses, and SBA does not track the business size of the applicant.

[37] Although our report focuses on business loans, approximately 83 percent of the original loan applications SBA received after Hurricane Sandy were home loan applications. (In the case of "reconsiderations"—that is, applications that were declined upon first submission and resubmitted for a second review—and "appeals"—that is, reconsidered applications that were declined and resubmitted for a third review—SBA includes the original application submission in the total of "original loan applications" but not the second or third

submissions). SBA received a total of 85,456 original disaster loan applications— 70,518 home loan applications and 14,938 business loan applications.

[38] The total approval rates for the previous disasters, in reverse chronological order, are Hurricane Irene, 37 percent; Hurricane Ike, 22 percent; Hurricane Wilma, 49 percent; Hurricane Rita, 37 percent; and Hurricane Katrina, 46 percent.

[39] SBA can cite more than one reason for declining a loan application. Of the 5,663 Hurricane Sandy applications that SBA denied, 1,020 (18 percent) had two or more decline codes applied.

[40] A withdrawn application is one that is removed from consideration before SBA has rendered a decision regarding approval or decline. Either SBA or an applicant can withdraw a business loan application. The 14,558 figure for total applications omits 380 applications that remained in various stages of processing as of January 31, 2014.

[41] Loan officers may assign multiple codes to each withdrawn application. However, this occurred in only 40 instances, representing less than 1 percent of all withdrawals.

[42] SBA disaster loans can be cancelled after approval. A cancellation does not impact the calculation of SBA's approval rate—approved loans are recorded as approvals even if subsequently cancelled. Either SBA or the borrower can initiate a cancellation. SBA loan officers assign codes to cancelled loans. Of the 15 cancellation codes, 7 are designated as "Agency Cancellations"; the remaining 8 codes are designated as "Cancellations at Borrower's Request."

[43] Unlike declined and withdrawn applications, to which SBA can assign multiple decline and withdrawal codes (respectively), each cancelled loan was assigned only one cancellation code.

[44] We did not ask entities we spoke with to provide information on reasons for cancellations.

[45] GAO, *Small Business Administration: Additional Steps Should Be Taken to Address Reforms to the Disaster Loan Program and Improve the Application Process for Future Disasters*, GAO-09-755 (Washington D.C.: July 29, 2009).

[46] According to SBA officials, the agency received $1,310,000 for administrative expenses; $352,357 for the IDAP subsidy; and $1,337,643 for the EDALP subsidy. As of August 2014, SBA officials stated that $400,000 in administrative expenses has been expended.

[47] GAO, *Small Business Administration: Progress Continues in Addressing Reforms to the Disaster Loan Program*, GAO-12-253T (Washington D.C.: Nov. 30, 2011).

[48] Section 7(a) of the Small Business Act, as amended, codified at 15 U.S.C. § 636(a). The 7(a) program is intended to serve creditworthy small business borrowers who cannot obtain credit through a conventional lender at reasonable terms and do not have the personal resources to provide financing themselves. Under the 7(a) program, SBA guarantees loans made by commercial lenders to small businesses for working capital and other general business purposes. E-Tran is SBA's electronic platform for loan guarantee originations and servicing. For more information see SBA's website, http://www.sba.gov/content/e-tran, accessed on August 25, 2014.

[49] SBA One is described as an improved lending platform that will simplify the lending process by using one set of forms for all 7(a) loans; serve as a one-stop shop for all steps of the loan process, from determining eligibility through closing out the loan; and provide one data management system to measure and evaluate loan trends and performance.

[50] If an applicant receives a direct disaster loan, the applicant is statutorily required to use the proceeds to repay the IDAP loan before using the proceeds for any other purpose. 15 U.S.C. § 657c(n).

[51] If an SBA-guaranteed loan defaults, the lender must enter the guarantee purchase process by requesting that SBA honor the guarantee. During this process, SBA reviews whether the lender has complied with the SBA loan authorization for the loan, SBA requirements, and prudent lending practices before making a decision to honor the guarantee. For more information on the guarantee purchase process, see SBA's website, http://www.sba.gov/category/lender-navigation/steps-sba-lending/7a-loans ase, accessed on August 7, 2014.

[52] If an applicant receives an IDAP loan before being declined for a direct disaster loan (through the Disaster Loan Program), the applicant is required to repay the loan not earlier than 10 years after the date of final disbursement 15 U.S.C. § 657n(d)(2). SBA regulations state that the maturity of an SBA IDAP loan must be at least 10 years from the date of final disbursement, but no more than 25 years. 13 C.F.R. § 123.703(d)(2).

[53] Lenders may charge a late payment fee not to exceed 5 percent of the scheduled IDAP loan payment, and reasonable direct costs for liquidation.

[54] GAO/AIMD-00-21.3.1

[55] GAO-09-755.

End Notes for Appendix I

[1] We obtained data from October 2012—when Hurricane Sandy made landfall—through January 2014 for all analyses for this report, unless otherwise noted. We selected January 2014 for the end date because over 99 percent of Hurricane Sandy loan applications had been processed by this date.

[2] Reconsidered loan applications are those that were previously declined once and subsequently resubmitted to SBA. Appealed loan applications are those that were previously declined twice and subsequently resubmitted to SBA. By including the average number of days it took to process these loan applications (in addition to originally submitted applications), we capture the total amount of time SBA spent on processing loan applications at each phase.

[3] SBDCs are SBA resource partners that provide an array of technical assistance to small businesses and entrepreneurs within their area. Further, some SBDCs serve as Business Recovery Centers that assist business owners during disasters.

[4] The selected counties in New York were Kings (Brooklyn), Nassau, Richmond (Staten Island), and Queens counties. The selected counties in New Jersey were Atlantic, Hudson, Monmouth, and Ocean counties.

[5] For these calculations, we used only final decisions on application files and did not include any intermediate decisions made before the final decision. For example, we counted a reconsidered loan that was later approved as a single approval rather than a decline and an approval.

In: Hurricane Sandy Disaster Relief
Editor: Edward S. Ramirez

ISBN: 978-1-63482-174-2

Chapter 3

Emergency Transportation Relief: Agencies Could Improve Collaboration Begun During Hurricane Sandy Response*

United States Government Accountability Office

Abbreviations

CBO	Congressional Budget Office
CRS	Congressional Research Service
DHS	Department of Homeland Security
DOT	Department of Transportation
DRAA	Disaster Relief Appropriations Act, 2013
DRF	Disaster Relief Fund
FEMA	Federal Emergency Management Agency
FHWA	Federal Highway Administration
FRA	Federal Railroad Administration
FTA	Federal Transit Administration
MAP-21	Moving Ahead for Progress in the 21st Century Act
NORTA	New Orleans Regional Transit Authority
OIG	Office of Inspector General

* This is an edited, reformatted and augmented version of The United States Government Accountability Office publication, GAO-14-512, dated May 2014.

WHY GAO DID THIS STUDY

In October 2012, Hurricane Sandy devastated portions of the Mid-Atlantic coast causing severe damage to transit facilities and infrastructure and disrupting mobility in the New York metropolitan region. In January 2013, the President signed the DRAA, which provided approximately $50.5 billion in federal aid for expenses related to Hurricane Sandy. GAO was asked to examine DRAA emergency relief assistance for transportation.

This report addresses (1) the progress DOT has made allocating, obligating, and disbursing DRAA surface transportation funds, (2) how FTA's new Public Transportation Emergency Relief program compares to FEMA's and FHWA's emergency relief programs, and (3) the extent to which FTA and FEMA have implemented their memorandum of agreement to coordinate their roles and responsibilities when providing assistance to transit agencies. GAO analyzed relevant laws, regulations, and agency documentation, and interviewed DOT, FEMA, and New Jersey and New York area transit officials.

WHAT GAO RECOMMENDS

GAO recommends that DOT and the Department of Homeland Security (DHS) direct FTA and FEMA to establish specific guidelines to monitor, evaluate, and report the results of collaborative efforts—including their communications program and protocol as contemplated in their memorandum of agreement. DHS agreed with our recommendation, and DOT took no position. DHS and DOT also provided technical comments, which we incorporated as appropriate.

WHAT GAO FOUND

The Department of Transportation (DOT) is in the process of allocating, obligating, and disbursing the $13 billion appropriated by the Disaster Relief Appropriations Act, 2013 (DRAA) for surface transportation relief. Most of the DRAA surface transportation funds—over $10 billion—were appropriated to the Federal Transit Administration's (FTA) new Public Transportation Emergency Relief Program. An FTA damage assessment in January 2013 estimated the costs of repairing facilities damaged by Hurricane Sandy in New York and New Jersey to be about $5.7 billion. To date, FTA has obligated about $1.5 billion for 15 grants and disbursed about $499 million to reimburse transit agencies for emergency response, recovery, and repair costs. These disbursements are consistent with Congressional Budget Office estimates, and transit projects can take years to complete. Furthermore, FTA plans to use nearly half of its DRAA appropriation for resiliency projects (or projects to protect facilities from future damage), most of which will be carried out through a competitive grant process. FTA was evaluating applications when GAO completed its review.

FTA's new Public Transportation Emergency Relief Program has more flexibility and fewer restrictions in funding projects compared to the Federal Emergency Management Agency's (FEMA) Public Assistance and Hazard Mitigation programs and the Federal Highway Administration's (FHWA) Emergency Relief Program. For example, FEMA's Hazard Mitigation program places limits on the amount of emergency relief funds that can be used for resiliency projects, while FTA's program does not. FTA's program also has more flexibility in how funds can be used for repairs, allowing transit agencies to improve facilities beyond pre-disaster conditions. The use of emergency relief funds for projects that go beyond recovery efforts is not new—activities funded by FHWA's Emergency Relief Program have also expanded beyond repair and reconstruction. The expanding scope of emergency relief assistance illustrates the fiscal exposure the federal government faces and the challenges of establishing long-term sustainable funding for disaster relief and recovery.

Although FTA and FEMA have a memorandum of agreement for assisting transit providers during emergencies, they are limited in their ability to delineate specific roles and responsibilities for future disasters. This limit is because while FEMA receives funding on an ongoing basis, FTA, to date, has only received a supplemental appropriation for Hurricane Sandy and does not know what resources it will have for future disasters. Because FTA and FEMA have the authority to fund many of the same activities by law, transit agencies may experience confusion when seeking assistance under some circumstances. FTA and FEMA have not determined how collaborative efforts, including their communications program and protocol contemplated in the memorandum of agreement, will be monitored, evaluated, and reported, but instead rely on informal communication. As GAO has previously concluded, creating a means to evaluate the results of collaborative efforts can enhance and sustain them, and informal communications between federal agencies do not ensure that collaboration is effective. Establishing more formal monitoring and evaluation of combined efforts could help FTA and FEMA ensure effective collaboration.

* * *

May 28, 2014

The Honorable Tom Latham
Chairman
Subcommittee on Transportation,
Housing and Urban Development, and Related Agencies
Committee on Appropriations
House of Representatives

Dear Mr. Chairman:

Public transportation agencies are vulnerable to natural and man-made disasters that can disrupt transit operations and destroy vehicles and facilities. This presents a range of challenges, as transportation infrastructure is a key resource before, during, and after a major disaster or emergency. Before a disaster, public transportation infrastructure and resources can be used to assist with evacuating residents. During a disaster, the loss of transit can worsen the disaster's impact by impeding a community's access to emergency relief services and medical care. After a disaster, public transportation can be used to return people to their homes by restarting normal operations and allowing people to return to their jobs and daily activities.

In October 2012, Hurricane Sandy hit the Mid-Atlantic coast during high tide, bringing with it a storm surge and flooding. The President issued disaster declarations under the Robert T. Stafford Disaster Relief and Emergency Assistance Act (Stafford Act)[1] for the following areas: Connecticut, Delaware, District of Columbia, Maryland, Massachusetts, New Hampshire, New Jersey, New York, Ohio, Pennsylvania, Rhode Island, Virginia, and West Virginia. Parts of New York and New Jersey were the hardest hit, and mobility in the greater New York City metropolitan area was temporarily crippled, including its transit systems, which normally carry more customers than all other systems in the United States combined. In January 2013, the President signed the Disaster Relief Appropriations Act, 2013 (DRAA), which provided approximately $50.5 billion in supplemental appropriations to 61 programs at 19 federal agencies for expenses related to Hurricane Sandy.[2] DRAA provided over $13 billion to the Department of Transportation (DOT), most of it to the Federal Transit Administration's (FTA) new Public Transportation Emergency Relief Program, which had gone into effect 28 days before Hurricane Sandy struck the United States. However, not all of the funds appropriated to FTA under DRAA were available until DOT signed a memorandum of agreement on March 4, 2013, with the Department of Homeland Security (DHS) to coordinate the roles and responsibilities of the two departments in providing assistance for public transportation.[3] The Federal Emergency Management Agency (FEMA), within DHS, is the lead agency responsible for preparing the federal response plans and programs for future disasters, coordinating the federal response when a disaster is declared, and providing grants to assist state and local governments during the response and recovery phases of a disaster. Within DOT, the Federal Highway Administration (FHWA) also has an Emergency Relief Program that provides funds for the repair and reconstruction of roads on the federal-aid highway system that have suffered serious damage as a result of either a natural disaster or a catastrophic failure from any external cause.

Given the amount of funds distributed and the number of agencies involved, you requested that we study emergency relief for transportation facilities and the services being carried out under the DRAA. We reviewed DOT's progress in disbursing DRAA funds to address Hurricane Sandy—with a focus on FTA's new Public Transportation Emergency Relief Program, how it compares to other federal agencies that provide emergency relief funding to surface transportation, and implications for responding to future disasters. Specifically, this report addresses (1) the progress DOT has made allocating, obligating, and disbursing DRAA surface transportation funds; (2) how FTA's new Public Transportation Emergency Relief Program compares to FEMA's and FHWA's emergency relief programs; and (3) the extent to which FTA and FEMA have implemented their memorandum of agreement to coordinate their roles and responsibilities when providing assistance to transit agencies affected by Hurricane Sandy or future disasters.

To address all objectives, we reviewed and analyzed relevant laws and regulations; our previous reports; FTA, FWHA, and FEMA guidance; and interviewed officials from DOT and FEMA. We interviewed officials from seven of the 13 transit agencies that received grants from FTA. We selected the transit agencies that received FTA's four largest funding allocations (all located within New York and New Jersey) and selected three agencies that received smaller funding allocations but were located outside of New York and New Jersey. Our findings from interviews with these selected agencies provide insights and illustrate FTA's emergency transit-relief grant process; however, the results of our review should not be used to make generalizations about all FTA emergency transit relief grants. We also interviewed officials from Amtrak,[4] which received funds from the Federal Railroad Administration (FRA) within DOT.

To determine the progress DOT has made with DRAA surface transportation funds, we obtained information from the DOT surface-transportation agencies—FTA, FHWA, and FRA—on the amount of DRAA funds that have been allocated, obligated, and disbursed from March 2013 to April 2014. We obtained FTA's damage assessment of public transit agencies in New York and New Jersey affected by Hurricane Sandy. We reviewed the Congressional Budget Office's (CBO) 2012 analysis of the DRAA. To assess the reliability of the allocation, obligation, and disbursement data we received from DOT, we reviewed relevant documentation, interviewed knowledgeable officials, and reviewed internal controls related to the handling of the data. We found the data to be sufficiently reliable for our purposes. To determine how FTA's Public Transportation Emergency Relief Program compares to FEMA's and FHWA's emergency relief programs, we reviewed and analyzed relevant statutes and regulations, as well as documents pertaining to FTA, FHWA, and FEMA emergency relief programs to determine similarities and differences between the programs' eligibility rules, requirements, and funding.[5] To determine the extent to which FTA and FEMA have coordinated their roles and responsibilities when providing disaster assistance to transit agencies, we reviewed and analyzed their memorandum of agreement and compared the agencies' efforts to practices that we have found can enhance collaboration among federal agencies.

BACKGROUND

We conducted this performance audit from April 2013 to May 2014 in accordance with generally accepted government auditing standards. Those standards require that we plan and perform the audit to obtain sufficient, appropriate evidence to provide a reasonable basis for our findings and conclusions based on our audit objectives. We believe that the evidence obtained provides a reasonable basis for our findings and conclusions based on our audit objectives. Appendix I provides a more detailed description of our scope and methodology.

When a disaster overwhelms the ability of state, local, and voluntary agencies to adequately provide essential services on their own, the federal government generally plays a central role in disaster response, providing selected resources where they are needed. The federal government has provided significant funds for transit services following past catastrophic disasters. For example, about $4.7 billion was provided in emergency supplemental appropriations for emergency transit services in response to the September 11, 2001, attacks and $232 million in response to the 2005 Gulf Coast hurricanes.

After Hurricane Sandy, Congress enacted and the President signed into law the DRAA, which provided supplemental appropriations to federal agencies to assist, improve, and streamline Hurricane Sandy recovery efforts, including efforts to address damage to public transportation systems. Figure 1 shows the funding provided by the DRAA by agency, including over $13 billion to DOT.[6]

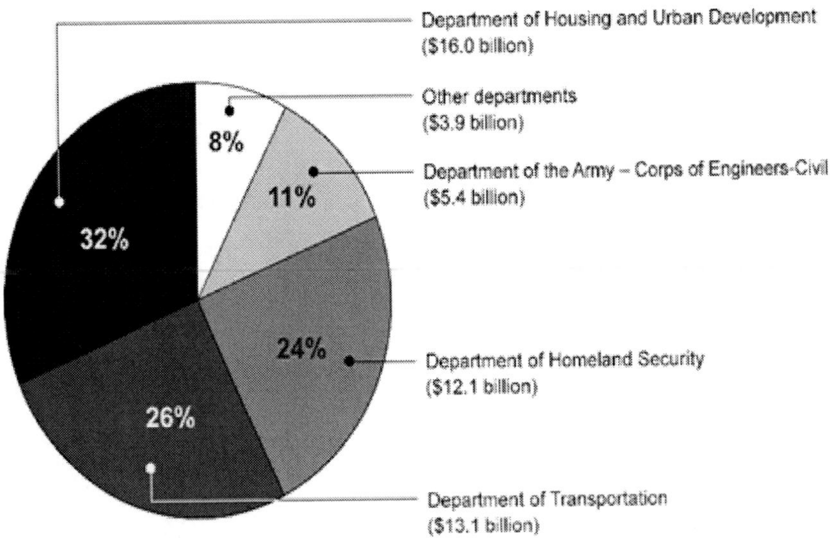

Source: GAO analysis of the Disaster Relief Appropriations Act, 2013.

Figure 1. $50.5 Million in DRAA Funding by Agency (As Enacted on January 29, 2013).

Most of DOT's DRAA funding was made available to assist transit agencies, especially in New York and New Jersey, which experienced substantial damage to their infrastructure and incurred significant costs to restore operational service after the storm. All of these transit systems experienced the temporary suspension of service and most sustained serious, widespread damage, which affected the region's entire transportation network. As alternative modes of travel were sought, the use of the region's roadway network far exceeded capacity. The ripple effects were felt by many residents of the region who were not able to travel.

Figure 2 below illustrates some of the damage to regional transportation systems as a result of Hurricane Sandy.

Sandbags at the entrances of Port Authority Trans-Hudson rail transit station in Hoboken, N.J.
Source: Port Authority of New York and New Jersey.

Damage to the Metropolitan Transit Authority Rockaway Line
Source: New York Metropolitan Transportation Authority (MTA)

Flood waters pouring into the Hoboken Station on the Port Authority Trans-Hudson (PATH) System

Damage to the Morgan Drawbridge on the New Jersey Transit System
Source: New Jersey Transit.

Source; Port Authority of New York and New Jersey

Figure 2. Examples of Damage from Hurricane Sandy.

FEMA is the federal government's primary agency for disaster response and has primary responsibility for administering the provisions of the Stafford Act and for carrying out activities for national preparedness. Specifically, FEMA (1) funds and coordinates eligible prevention, protection, and mitigation activities; (2) provides and coordinates the immediate federal response to save lives and property; (3) funds the reconstruction of disaster-damaged public infrastructure; and (4) provides individual assistance to stricken families. In addition to coordinating disaster response and recovery operations, FEMA's Public Assistance Program provides funding to state and local governments and some nonprofit organizations for recovery efforts after a disaster, including removing debris, implementing emergency protective measures, and repairing or replacing damaged public equipment or facilities. FEMA also provides grants to states and local governments to implement mitigation measures that reduce or permanently eliminate future risk to lives and property from natural hazards under its Hazard Mitigation Grant Program. This program funds projects in accordance with priorities identified in state, tribal, or local hazard mitigation plans and enables mitigation measures to be implemented during the immediate recovery from a disaster.

DOT also has two emergency relief programs to assist with damage to surface transportation infrastructure and facilities after a disaster, both of which received a share of DOT's DRAA funding for Hurricane Sandy.

- FHWA's Emergency Relief Program provides funding for emergency and permanent repairs or reconstruction of roadways on the federal-aid highway system and on federal lands that have suffered serious damage as a result of a natural disaster or catastrophic failure from an external cause. Emergency repairs are completed during and immediately following a disaster to quickly restore essential highway traffic service and protect remaining facilities, for example, establishing emergency detours and providing temporary bridges or ferry service. Permanent repairs are undertaken after emergency repairs have been completed to restore damaged facilities to pre-disaster conditions, for example, restoring pavement surfaces or reconstructing damaged bridges. FHWA only provides funding under its Emergency Relief Program for highways that are either federal-aid highways or on federal lands. FEMA's Public Assistance Program provides funding to repair disaster damage to other roads.
- FTA's Public Transportation Emergency Relief Program was authorized by the Moving Ahead for Progress in the 21st Century Act (MAP-21).[7] MAP-21 provides FTA with the primary responsibility for reimbursing emergency response and recovery costs after an emergency or major disaster affects a public transportation system. The FTA program is a reimbursable grant program and allows FTA to make grants for capital projects to protect, repair, reconstruct, or replace equipment and facilities of a public transportation system as well as for eligible operating costs. Such costs include reestablishing,

expanding, or relocating public transportation route service in the event of a natural disaster that affects a wide area or a catastrophic failure from any external cause.

Most of the $13 billion that DRAA made available to DOT for surface transportation relief was appropriated to FTA for the Public Transportation Emergency Relief Program.[8] The DRAA provides this funding for eligible transit operating expenses, capital projects to repair or reconstruct damaged facilities, and projects to protect facilities from future damage (resiliency projects). According to FTA's interim program regulations, resiliency projects are designed and built to address a transit system's future vulnerabilities due to emergencies or major disasters that have occurred and are likely to reoccur in the geographic area in which the public transportation system is located, or projected changes in development patterns, demographics, or extreme weather or other climate patterns.[9] For example, MTA is planning a resiliency project to increase New York City's pumping capacity by converting existing rolling stock into pump trains. The project is intended to shorten the period of time it takes to pump water out of the subway system to prevent major damage from saltwater infiltration into the subway system. The DRAA also provided the DOT Secretary authority to transfer up to $5.4 billion for resiliency efforts to other DOT agencies, such as FHWA and FRA.[10]

In January 2013, FTA, FEMA, and affected transit agencies worked together after Hurricane Sandy to assess the damage in the New York and New Jersey region. The damage assessment estimated that the damage to public transit was about $5.7 billion. About 99 percent of the estimated damage was sustained by four transit agencies, as shown in table 1.

Table 1. Estimated Damage to New York and New Jersey Transit Agencies, As Assessed by FTA, FEMA, and Transit Agencies (January 31, 2013)

Agencies	Restoration/Recovery
New York Metropolitan Transportation Authority (MTA)	3,904
Agencies	Restoration/Recovery
Port Authority of New York/New Jersey	1,286
New Jersey Transit	459
New York City DOT	36
Other New York/New Jersey Agencies	4
TOTAL	$5,689

Source: FTA.

FTA officials said that the damage assessment only covered damage in New York and New Jersey because this was the area with the most extensive damage.

Rather than identify total damage in other areas, FTA established a process for potential recipients to identify their damage and to apply for funds. FTA made initial grants to recipients outside of New York and New Jersey, but those grants account for less than 1 percent of all allocations made as of April 30, 2014.

For example, FTA awarded a $5,352 grant to the Milford Transit District (MTD) in Connecticut for evacuation services performed during Hurricane Sandy, transportation after the storm, and debris cleanup. Beyond that, FTA has not estimated an amount of repair and recovery outside of New York and New Jersey but has set aside funds for future allocations (approximately $28 million) as needed if there are eligible response and recovery expenses from agencies that may not have received pro-rated allocations.

FTA Has Allocated over Half of DOT's DRAA Funds While Setting Up the Public Transportation Emergency Relief Program

As figure 3 shows, the approximately $13 billion that DRAA appropriated to DOT for surface transportation relief was divided among several DOT agencies including FTA, FHWA, and FRA—but the vast majority of it was appropriated to FTA.

Two events have affected the amount of funds available to the DOT agencies.

First, all DRAA funds were reduced by 5 percent—from approximately $13 billion to almost $12.4 billion—on March 1, 2013, as a result of sequestration.[11]

Second, the DOT Secretary exercised his transfer authority under DRAA and approved a transfer in May 2013 of $185 million from FTA to FRA. This transfer was used to award a grant for Amtrak to install a concrete casing that preserves the right of way in the New York City Hudson Yards for the potential alignment of a future trans-Hudson tunnel.

This concrete casing would allow access for a future project to install an additional railway tunnel under the Hudson River and increase train capacity for commuter, regional, and long distance intercity services going in and out of New York City (see fig. 4). The Gateway Tunnel project has been planned since 2010, and with this funding, Amtrak is able to preserve a possible right-of-way for the project. DOT officials told us that there are no plans at this time to make any other transfers using this authority.

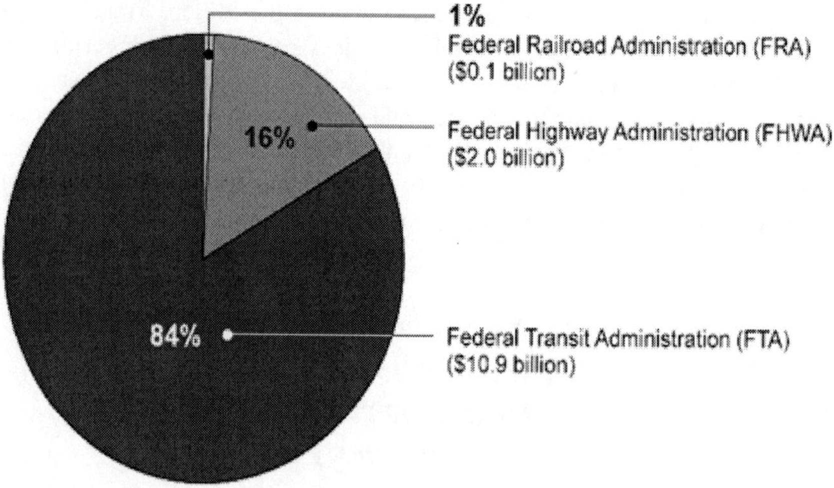

Source: GAO analysis of the Disaster Relief Appropriations Act, 2013.
Note: Percentages do not add to 100 due to rounding.

Figure 3. DRAA Appropriations for Surface Transportation (As Enacted on January 29, 2013).

DOT surface transportation agencies have allocated about $6.3 billion (51 percent), obligated about $2.2 billion (17 percent), and disbursed about $721 million (6 percent) of the roughly $12.4 billion available DRAA funds for transportation relief as of April 30, 2014.[12]

The available funds for each agency shown in table 2 reflect the impact of sequestration, and the DOT Secretary's transfer from FTA to FRA.

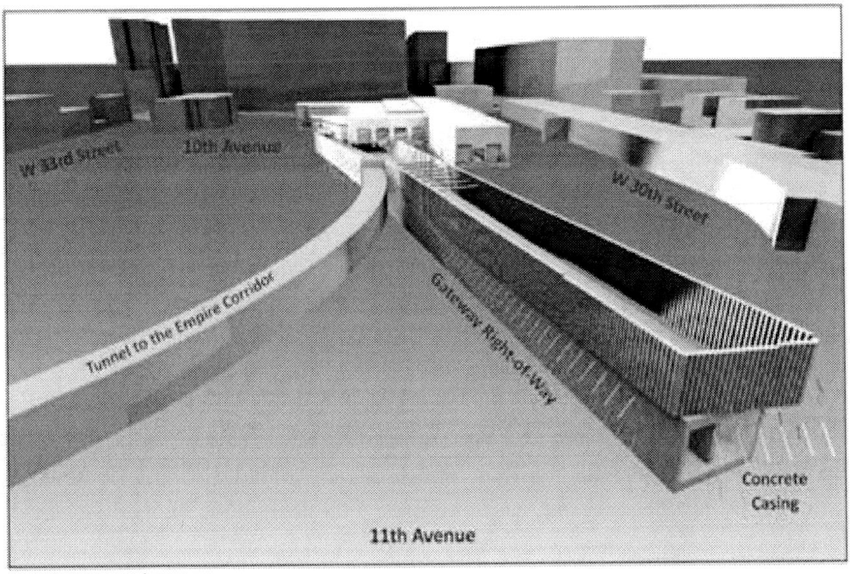

Source: Amtrak.

Figure 4. Depiction of the Amtrak Gateway Tunnel Project between New York and New Jersey.

Table 2. DOT DRAA Funding Status ($ millions) as of April 30, 2014

DOT Surface Transportation Administrations	Available funds	Allocated	Obligated	Disbursed
Federal Highway Administration	1,921	616	473	120
Federal Railroad Administration	297	a	215	102
Federal Transit Administration	10,170	5,679	1,463	499
TOTAL	$12,388	$6,295	$2,151	$721

Source: DOT.

[a] FRA funds are not allocated to grantees as they are for FHWA and FTA programs; FRA grants are awarded from available funds.

FTA Progress

FTA is in the process of allocating and distributing DRAA funds while facing unique challenges setting up its Public Transportation Emergency Relief Program. The program was new when Hurricane Sandy struck, and

FTA received its first funds for the program through the DRAA just 3 months after the storm. As a result, as of April 2014, FTA was finalizing its program rules at the same time it was distributing billions of dollars of DRAA funds to affected transit agencies. Furthermore, the amount of funds that FTA is distributing through the emergency relief program (over $10 billion) is almost as much as FTA's total budgetary resources were in fiscal year 2012 (approximately $12.1 billion).

FTA's over $10 billion in DRAA funding is more than enough to address the costs of repairing transit facilities damaged by Hurricane Sandy in New York and New Jersey (see table 1). The January 2013 damage assessment estimated Hurricane Sandy damage in New York and New Jersey to public transit to be about $5.7 billion, but also identified about $10 billion in potential resiliency projects to guard against future disasters. As of April 30, 2014, FTA has allocated almost $5.7 billion of its DRAA funding to transit agencies affected by Hurricane Sandy and obligated a total of about $1.5 billion for 15 grants for emergency response, recovery, repair and reconstruction activities.[13] When we completed our review, transit agency officials we spoke with were generally positive about the FTA Public Transportation Emergency Relief Program and told us that, in their experience, FTA has not caused them any delays in receiving funding. Of the amount allocated, FTA made $1.3 billion available for what it called "locally prioritized" resiliency funding. This funding is primarily intended for resiliency improvements completed in tandem with recovery and rebuilding projects that will improve cost effectiveness and can be implemented quickly.

As of April 30, 2014, FTA has only disbursed $499 million of its DRAA funds, but this does not indicate a lack of progress for several reasons. FTA's progress is in line with expenditure estimates. In December 2012, CBO estimated that only 3 percent of FTA's DRAA funds would be expended in the first year and forecasted that it would take over a decade for FTA to disburse all the funds based on historical patterns.[14] As of January 29, 2014—one year after DRAA was enacted—FTA had disbursed almost 3 percent of its DRAA funds, approaching CBO estimates. In addition, transit projects are long-term investments that typically take years to complete. For example, the MTA Long Island Rail Road's West Side Storage Yard—which undertakes inspections, repairs, maintenance, and cleaning for the railroad's fleet—suffered severe damage due to the tidal surge and resulting intrusion of salt water during Hurricane Sandy. The project to repair it will replace numerous assets in the yard, including switch machines, signal components, third-rail components, switch heaters, and fire alarm systems. According to officials, MTA plans to complete this work in phases and estimates a February 2018 completion. Finally, FTA plans to use about 42 percent of its DRAA appropriation for resiliency projects, most of which will be carried out through a competitive grant selection process, and FTA was evaluating applications when we completed our review.

Specifically, in December 2013, FTA issued a Notice of Funding Availability for approximately $3 billion in funding to be awarded for resiliency projects on a competitive basis.[15] This funding differs from the $1.3 billion that has been obligated for locally prioritized resiliency funding. Any transportation agency[16] within the declared Hurricane Sandy disaster area can apply for these funds,[17] and they can be used to protect either a facility or system that has been repaired or replaced after Hurricane Sandy, or that is at risk of being damaged or destroyed by a future emergency or natural disaster.[18] Among the other criteria, FTA will evaluate and award competitive resiliency funding based on the cost-effectiveness of the project, implementation strategy, extent to which improvements will protect vulnerable existing infrastructure, availability of local financial commitments, and the technical capacity of grantees. According to FTA's notice, in developing FTA guidelines for resiliency projects, it considered recommendations made by the Hurricane Sandy Rebuilding Task Force.[19] The Task Force specifically recommended that infrastructure rebuilt with DRAA funds be designed to increase the resilience of the region.[20]

The DRAA appropriation to FTA provided funding for the DOT Office of Inspector General (OIG) to conduct oversight of FTA's DRAA funds. In the initial report, the OIG stated that FTA's initial response to Hurricane Sandy was noteworthy, but that FTA has yet to fully address challenges in its oversight plans and procedures.[21] The OIG made several recommendations to FTA, including specifying the required time frames for completing risk assessments for all grantees and finalizing program guidance that incorporates lessons learned from emergencies and best practices from emergency relief guidance. According to the OIG's report, FTA has provided appropriate planned actions and timeframes for completion for most of the OIG recommendations.

FTA's Relief Program Is More Flexible Than FEMA's and FHWA's but Has Not Received Ongoing Funding

Compared to FEMA's Public Assistance and Hazard Mitigation Grant Programs and FHWA's Emergency Relief Program, FTA's new Public Transportation Emergency Relief Program has more flexibility and fewer restrictions in funding emergency relief projects. For example, FEMA's Hazard Mitigation Grant Program places limits on the amount of emergency relief funding that can be allocated toward resiliency projects, whereas FTA's program does not. However, unlike FEMA's and FHWA's programs, FTA's program has not received funding on an ongoing basis, which could limit some transit providers' ability to respond immediately after a disaster.

FTA's Program Has More Flexibility and Fewer Restrictions

FTA's Public Transportation Emergency Relief Program provides funding for similar types of activities as FEMA's and FHWA's emergency relief programs, such as emergency protective measures, permanent repairs, and resiliency projects, but the legislation authorizing FTA's program placed no limits on the amount that can be awarded for resiliency projects. Based on this flexibility, FTA is awarding about 42 percent of the funding it received for Hurricane Sandy, $4.3 billion, for resiliency projects.

In contrast, FEMA has limits on the percentage of emergency funding that may be used for resiliency projects. FEMA's Hazard Mitigation Grant Program provides funding for resiliency projects (referred to as hazard mitigation projects) but the amount of funding available for this program for a particular disaster declaration is limited. For example, under this program a state may be eligible to receive up to 15 percent of the estimated total of federal assistance for hazard mitigation grant assistance. States that meet higher mitigation planning criteria may qualify for up to 20 percent of assistance.[22] Of the $119.7 billion that FEMA has obligated for disaster assistance over a 10-year period—from 2004 through 2013—FEMA has obligated $4.9 billion (about 4 percent) for its Hazard Mitigation Grant Program activities to prevent or ease the impact of natural disasters.

Besides having greater flexibility in funding resiliency projects, FTA's Public Transportation Emergency Relief Program has fewer restrictions and greater flexibility to fund emergency relief projects compared to FHWA's and FEMA's programs, as discussed below:

FTA's program funds have more flexibility in how they can be used for repairs. FTA's Public Transportation Emergency Relief Program allows transit agencies to use response and recovery awards to improve facilities beyond pre-existing conditions by 1) replacing damaged assets with new assets that incorporate current design standards, 2) replacing obsolete equipment, or 3) bringing assets to a state of good repair as part of its recovery effort. By comparison, FEMA's Public Assistance Program will generally provide funding for permanent repairs to restore a facility based on the adopted codes and standards to its pre-disaster design, function, and capacity.[23] FEMA's Public Assistance Program may provide funding to replace the facility to current applicable codes and standards, if the grantee can demonstrate that the estimated repair cost exceeds 50 percent of the estimated replacement cost. FEMA's Public Assistance Program also provides some funding to be used for hazard mitigation measures. This funding, along with FEMA's Hazard Mitigation Program funding, can be used to improve facilities beyond pre-disaster conditions if applicants can demonstrate that the project is cost-effective. Like FTA's program, FHWA's Emergency Relief Program will also provide emergency relief funding to restore or replace a facility or bridge to current geometric and construction standards. However, FHWA limits this funding only to portions of the facility that are damaged and requires applicants to show that the improvement is cost-effective, prevents future recurring damage, or is technically feasible. In addition, FTA has more flexibility in how it can support emergency transit operations. For example, FTA's emergency relief program provides funding for emergency operations such as evacuations; rescue operations; ferry service to replace inoperable rail service or to detour around damaged areas; returning evacuees to their home after Hurricane Sandy; and for the net project costs related to reestablishing, expanding, or relocating public transportation service before, during or after Hurricane Sandy. By comparison, under FEMA's Public Assistance Program, financial assistance for transit and transportation operations are limited to those operations that are necessary to save lives and protect public health, such as evacuation or transportation of evacuees to emergency shelter. According to FEMA officials, FEMA does not have the authority to reimburse state or local transit agencies for expenses associated with emergency transit operations.

FTA emergency relief funds have no restrictions on funding projects included in statewide transportation improvement programs. FTA emergency relief funds can be used to fund projects that are included in statewide transportation improvement programs. In comparison, under MAP-21, FHWA emergency relief funds may not be used to repair or reconstruct a bridge if the construction phase of a replacement structure is already included in the approved statewide transportation improvement program at the time of an eligible event. Similarly, according to FEMA's Pubic Assistance Program guidance, funding may not be used to replace a facility if the facility is already scheduled to be replaced using federal funds, and work is scheduled to begin within 12 months of the time the disaster struck. According to FTA officials, funding for projects in statewide transportation improvement programs are not limited because FTA does not believe such plans constitute a firm commitment of funding. However, projects for which funds were obligated in an FTA grant prior to the declared emergency or disaster are not eligible to receive FTA emergency relief funds.[24]

FTA's cost-sharing arrangements and rules for non-federal sharing commitments are less restrictive. As shown in table 3, cost-sharing arrangements vary among FTA's, FEMA's, and FHWA's emergency relief programs.

Under FTA's program, for example, the federal share for capital and operating costs is generally up to 80 percent, and the nonfederal share is generally not less than 20 percent of the project cost. However, the DOT Secretary is authorized to waive in whole or in part FTA's non-federal share. Pursuant to the Stafford Act, the cost-sharing arrangements for FEMA's Public Assistance Program can also be adjusted, but only by the President at a governor's request.

Table 3. Federal and Non-Federal Cost-Sharing Arrangements for FTA's, FEMA's, and FHWA's Emergency Relief Programs

Agency	Program	Federal share	Non-federal share
FTA	Public Transportation Emergency Relief Program	Up to 80 percent of the project cost.[a]	Not less than 20 percent of the project cost.
FEMA	Public Assistance Program	Not less than 75 percent of the eligible project cost.[b]	Not more than 25 percent of the projectcost.
	Hazard Mitigation Program	Up to 75 percent of the eligible activity costs.	Not less than 25 percent of eligible activity costs.

Agency	Program	Federal share	Non-federal share
FHWA	Emergency Relief Program	Up to 100 percent for emergency repairs completed within 180 days after a disaster.[c] For permanent repair costs and emergency repair costs incurred after the first 180 days, the federal share is based on the type of federal-aid highway being repaired.[d]	Varies depending on the type of federalaid highway.

Source: GAO analysis of FTA, FEMA, and FHWA information.

[a] The DOT Secretary is authorized to adjust cost shares for the program and to waive in whole or in part the non-federal share requirement.

[b] Sections 403, 406, 407, and 502 of the Stafford Act provide that the Federal share of assistance shall be not less than 75 percent of the cost of the eligible work. FEMA regulations at section 206.47 of title 44, C.F.R., allow FEMA to recommend an increased cost-share. For example, in response to Hurricane Sandy, for the state of New York, the President authorized federal funds for all categories of the Public Assistance Program at 90 percent of total eligible costs, except assistance previously designated at 100 percent Federal share. In addition, FEMA recommends raising the usual 75 percent federal share for the Public Assistance Program to 90 percent when federal obligations, excluding FEMA administrative costs, meet a qualifying threshold.

[c] Emergency repairs completed within 180 days after a disaster to restore essential travel, minimize the extent of damage, or protect the remaining facilities are eligible for a 100 percent federal share.

[d] For example, for interstate highways, the federal share is 90 percent. For federal-aid facilities that meet the definition of a federal land access transportation facility, the federal share is 100 percent and for all other highways, the federal share is 80 percent. In addition, MAP-21 provided authority for the Secretary of Transportation to extend the 180-day period for 100 percent federal share of emergency repair work if there is a delay in the state's ability to access the site.

In addition, FTA's program permits recipients to waive local matching requirements by applying toll credits toward the non-federal match. Recipients of FHWA's Emergency Relief Program funds are specifically prohibited from using toll credits to reduce or eliminate the non-federal share.[25] States and localities that receive federal formula funds for highway or transit projects may apply "toll credits" to reduce or eliminate the funding they are required to provide to match federal funds for projects. These toll credits are earned when the state, a toll authority, or a private entity funds a capital transportation investment with toll revenues earned from existing toll facilities. FEMA's

Public Assistance and Hazard Mitigation Programs also do not allow for applicants to apply toll credits toward the local match but have other ways of meeting the non-federal share amount, for example, through in-kind contributions.[26]

FTA's program does not require a presidential declaration of disaster to release emergency relief funds. Under both FTA's and FHWA's programs, for a natural disaster or other event to be eligible for emergency relief funding, either the governor of a state has to declare an emergency and the Secretary of Transportation has to concur, or the President has to declare a major disaster under section 401 of the Stafford Act. However, a Presidential Declaration of a Major Disaster or Emergency is required for funding to be available for FEMA's Public Assistance Program. Likewise, FEMA's Hazard Mitigation Program requires a presidential declaration of a major disaster for funding to be available for that program.

FTA's program has flexibility to provide funds directly to the grantee. Under FEMA's and FHWA's programs, funding is distributed to the states. FEMA provides funding under its Public Assistance Program to eligible applicants through state and tribal government grantees. In our prior work, we found instances where funding from state grantees to transit agencies was delayed.[27] FHWA's Emergency Relief Program similarly provides funding to states, Puerto Rico, the District of Columbia, and territories. The program is administered through the state DOTs in close coordination with FHWA's division offices in each state. In contrast, FTA funding can be provided directly to transit agencies. Officials from the grantee agencies we spoke to told us that working with FTA is easier than working with FEMA, in part because of their ongoing grantee relationship with FTA and because FTA's personnel are transit specialists and its processes are better suited for complicated transit projects.

Recent Changes Have Expanded the Scope of Activities FHWA and FEMA May Fund

Even though FTA has more flexibility, particularly in the amount of funding it can allocate to resiliency projects, several recent changes to FHWA and FEMA program requirements have further expanded the scope of activities eligible for funding under these programs.

- MAP-21 expanded FHWA's scope of activities to allow its emergency relief funds to be used to expand the capacity of a structure—such as bridges, highways, or roads—being repaired or replaced to accommodate the volume of traffic that the structure will carry over its design life.[28] Prior to MAP-21, FHWA's Emergency Relief Program could only be used to expand the capacity of bridges to accommodate future traffic.
- The Sandy Recovery Improvement Act of 2013[29] revised the Stafford Act, establishing a new set of alternative procedures for FEMA's Public Assistance Program, allowing applicants to receive funding based on fixed estimates, rather than being reimbursed for actual costs. Under these new procedures, applicants can use grant funding that exceeds actual costs toward mitigation activities.

Use of emergency relief funds for projects that go beyond recovery efforts is not new. In our 2007 report on FHWA's Emergency Relief Program, we found that FHWA's use of Emergency Relief Program funds was not always limited to rebuilding facilities back to original conditions—the original intent of the program—and were used for projects that had grown in size and scope in response to public and community concerns. [30] For example, emergency relief funds were used to fund the reconstruction of the U.S. Highway 90 Biloxi Bay Bridge in Mississippi, a four-lane drawbridge between the cities of Biloxi and Ocean Springs, Mississippi. According to the Mississippi DOT officials, DOT initially proposed replacing the drawbridge with one that would provide an 85-foot clearance above Biloxi Bay. During a public comment period on the proposed bridge design, a local shipbuilder expressed concern that the height was not sufficient to allow for future ships to pass under the bridge. Mississippi DOT revised its proposed bridge design to provide a 95-foot clearance, which increased the cost of the bridge from an estimated $275 million to $339 million. We have also found that Congress has, on occasion, expanded FHWA's and FEMA's emergency relief eligibility criteria, an expansion that has resulted in projects receiving funding beyond what FHWA's and FEMA's emergency relief programs would otherwise have funded.[31] For example, in 2003, following the September 11, 2001, terrorist attack, Congress appropriated supplemental funding expanding FEMA's authority to allow it to fund recovery efforts and rebuild infrastructure beyond pre-disaster conditions to substantially improve commuter mobility.

While we have previously raised concerns about the expanding scope of activities eligible for emergency relief funding, we have also noted that building resiliency into infrastructure planning is important.[32] Enhancing the resilience of physical infrastructure, such as roads and bridges, vulnerable to extreme weather conditions can help reduce future economic losses and result in more efficient infrastructure planning. Federally funded road and bridge projects are typically expensive longterm investments. We have previously concluded that as the nation makes these investments, it faces the choice of paying more now, or potentially paying a much larger premium later to repair, modify, or replace infrastructure ill suited for future conditions.[33]

FTA's Public Transportation Emergency Relief Program Does Not Currently Receive Funding on an Ongoing Basis

FEMA and FHWA have funds available to provide recovery assistance when a disaster or emergency occurs:

- Funding for FEMA's Public Assistance and Hazard Mitigation Programs is available from the Disaster Relief Fund (DRF) after the President declares a disaster under the Stafford Act.[34] The DRF receives emergency-designated funding through both the regular appropriations process and through supplemental appropriations. The amount requested each year for the DRF in the President's Budget Request is based on the use of the fund (amount obligated from the fund) in prior years.[35] Once funds are appropriated to the DRF, they are available until expended, which means they carry forward from one year to another.
- Budget authority for FHWA's Emergency Relief Program is provided through contract authority, which authorizes FHWA to make commitments in advance of an appropriation.[36] FHWA is authorized to obligate up to $100 million each fiscal year for the program. Any unobligated balance remains available until expended.[37] In recent years, annual demands on FHWA's Emergency Relief Program have exceeded the $100 million annual authorization. For example, on average, the program's needed allocations for ordinary events—disaster events requiring under $100 million in federal funding—are 2.7 times the annual authorization, or $271 million. As a result, from 2005 to 2011, FHWA received about $7 billion in additional funding from general revenues through supplemental appropriations.

To date, Congress has not provided an annual appropriation for FTA's Public Transportation Emergency Relief Program, and FTA must rely on Congress to appropriate funds for a specific event. FTA's program has only received funds that Congress made available exclusively for Hurricane Sandy through the DRAA. Since it takes time for appropriations to be enacted, FTA is unable to release funding immediately after a disaster to transit providers which, in turn, could limit the ability of transit agencies that do not have resources to respond immediately after a disaster. However, annual budgeting for FTA's emergency relief program through regular appropriations would involve trade-offs with other funding priorities in a challenging budget environment. Events requiring the use of FEMA and FHWA emergency funds generally occur every year; however, transit disasters requiring FTA emergency relief funding may occur with less frequency. Ultimately, deciding how to allocate funding among competing demands is a policy decision for the Congress.

We discussed transit-funding challenges in our 2008 report, which highlighted the impact of challenges on the timeliness and effectiveness of FEMA's and FTA's assistance to transit agencies following the 2005 Gulf Coast hurricanes.[38] We found that neither FEMA nor FTA had mechanisms in place to provide transit funding immediately after the disasters. Transit officials from the New Orleans Regional Transit Authority (NORTA) told us that a lack of financial resources immediately after Hurricane Katrina delayed their disaster response efforts. Based on this work, we presented a range of possible options for Congress to consider for providing immediate financial assistance to transit agencies in response to a disaster, including that Congress create an emergency relief program in FTA for transit assistance modeled after FHWA's program with a "quick release mechanism"—a mechanism used to approve and release emergency funds within 1-2 days to begin the immediate flow of funds.[39] FHWA's program provides for a quick release of funding for larger disasters, wherein extensive damage is readily evident and state and federal level agencies need funds to flow quickly to the state. However, as we previously stated, to date, FTA has not received funding. If FTA's emergency relief program continues to receive only supplemental appropriations, FTA will not be able to quickly release funding for transit operations during and immediately after a disaster occurs.[40]

Long-standing Challenges of Identifying Long-term Sustainable Funding for Emergency Relief Efforts Remain

The expanding scope of emergency relief assistance and FTA's funding issues illustrate the long-standing challenges of establishing long-term sustainable funding for disaster relief and recovery.

- In 2013, we concluded that the federal government does not fully budget for emergency response and recovery costs and runs the risk of facing a large fiscal exposure at any time.[41] Over recent decades, the scope of emergency relief assistance has expanded, with disaster declarations increasing to a record of 98 in fiscal year 2011, compared with 65 in 2004, resulting in increased federal disaster costs. Over that period, FEMA obligated over $80 billion in federal assistance for disasters.[42] The federal government's fiscal exposure will only increase with the financial risks posed by extreme weather events. For this and other reasons, limiting the federal government's fiscal exposure by better managing climate change risks was added to GAO's High-Risk List.
- In our 2007 review of the FHWA Emergency Relief Program, we cited concerns about the use of emergency relief funds on an expanding scope of activities, noting that this gradual expansion of eligibility criteria was exacerbating the problem of longer-term sustainability.[43] We found that monies authorized from the Highway Trust Fund were not sufficient to cover the costs of ordinary events states experienced and that since 1990, the program had relied on supplemental appropriations to make up for the funding shortfall. We recommended that to put the Emergency Relief Program on a sound financial footing, additional alternatives to address the fiscal imbalance needed to be considered and that Congress could consider tightening the eligibility criteria for Emergency Relief funding, either through amending the purpose of the Emergency Relief Program, or by directing FHWA to revise its program regulations. We also recommended that Congress consider the expected future demands of the program and reexamine the appropriate level and sources of funding to address disasters—including whether to increase the $100- million annual authorized funding and whether the Highway Trust Fund, general revenues, or some combination would allow the program to accomplish its purpose in a fiscally sustainable matter.

- In another 2008 report, we reviewed supplemental emergency appropriations from fiscal years 1997–2006 and stated that to the extent possible, funds should be provided through the regular appropriations process to ensure that trade-offs are made among competing priorities, especially in an environment of increasingly constrained resources.[44]

Others have also commented on the process of funding emergency relief efforts in a fiscally sustainable manner. In its 2010 Report, the National Commission on Fiscal Responsibility and Reform[45] recommended that Congress explicitly set aside funds for disaster relief to better prepare for disasters. In a 2003 report, the Congressional Research Service (CRS) cited various challenges and benefits to having an emergency relief fund.[46] For example, according to CRS, maintaining emergency funds at a level that more accurately reflects emergency needs instead of relying on supplemental funding would foster fiscal responsibility and help ensure disaster response funding is budgeted with existing funds. However, CRS noted that, because disasters cannot always be anticipated, emergency supplemental appropriations can facilitate funding recovery efforts for disasters, the timing and severity of which are unpredictable.

FTA and FEMA Have Limited Ability to Assign Responsibilities in Their Memorandum of Agreement but Could Do More to Ensure Effective Collaboration

FTA's and FEMA's Ability to Clarify Specific Responsibilities for Future Disasters Is Limited

On March 4, 2013, FTA and FEMA signed a memorandum of agreement, as required by MAP-21, to coordinate their roles and responsibilities when providing assistance for public transportation in areas in which the President has declared a major disaster or emergency. We have found that by working together to define roles and responsibilities in collaborative efforts, federal agencies can clarify understanding about who will do what, organize their joint and individual efforts, and facilitate decision making.[47] However, in this memorandum of agreement, FTA and FEMA have limited ability to clarify who will do what related to their respective responsibilities when providing emergency transportation relief in future disasters. This is because, as we stated previously, FTA lacks the funding to assume costs for future disasters unless and until Congress provides an appropriation, either on an ongoing basis or for a specific disaster. Lacking information on the resources it will have, FTA is limited in pre-determining what assistance it could provide during future disasters. Conversely, because FHWA receives funding on an ongoing basis, FHWA and FEMA can delineate specific roles and responsibilities during emergencies that affect highways. FHWA provides funds to repair and replace disaster-damaged roads that are on the federal-aid highway system, and FEMA provides funds for roads that are not part of the system. However, FTA and FEMA are limited to more general descriptions of how they will cooperate—such as the principles and general areas of cooperation described in the memorandum of agreement. FEMA officials told us that after FTA finalizes all of its program rules, they may be able to make some small clarifications to their roles and responsibilities, but because of the uncertainty of FTA funding, they will not be able to draw clear distinctions in responsibilities, as they have been able to do with FHWA.

As a result, FTA and FEMA will have to determine their specific roles and responsibilities on a per-incident basis. Funds provided for the FTA Public Transportation Emergency Relief Program do not affect the ability of any other agency to provide any other funds authorized by law. Thus, FEMA and FTA can fund many of the same activities.[48] The memorandum of agreement states that if FTA receives funds, it will be the "primary" payor of expenses incurred by public transportation agencies as a result of emergency or major disaster. For Hurricane Sandy, FTA and FEMA agreed that FTA would assume this responsibility for affected transit agencies, because it received an appropriation from the DRAA large enough to cover the damage without FEMA funding. However, outside of Hurricane Sandy, FTA has been unable to assist transit agencies at all. For example, a transit agency in Massachusetts that had received assistance after Hurricane Sandy from FTA also sustained damage related to Winter Storm Nemo but went to FEMA for assistance instead.[49]

Although FTA had the authority to respond, and the transit agency official said that he preferred to go to FTA, FTA had no funding for any event outside of Hurricane Sandy and could not assist. In the future, agencies may continue to receive transit funding from FTA in one disaster and from FEMA in another. However, if a significant transit disaster occurs in the future and FTA funds are not enough to address the damage, transit agencies could go to FTA to fund some of their repairs and to FEMA to fund others, which could increase confusion and the chance that similar activities could be funded by both agencies inadvertently.

Furthermore, if FTA ultimately does receive supplemental funding after a specific emergency or disaster, transit agencies that receive assistance may have to change their application for emergency funds mid-stream from FEMA to FTA. In the immediate aftermath of Hurricane Sandy, FTA and FEMA officials were uncertain as to which agency would ultimately fund recovery efforts for transit. Transit agency officials affected by the storm said that at first, they were using FEMA worksheets to document their expenses. However, 3 months after the storm—after FTA received an appropriation under DRAA and became the primary payor of expenses—transit agencies had to re-do their paperwork. For example, one transit agency official told us that he put together a lot of documentation after Hurricane Sandy to receive funds from FEMA, documentation that was not necessary when the process changed to FTA. Another transit agency official told us that the requirements for FEMA's and FTA's programs are different and as a result, the agency had to make a change from ensuring FEMA's requirements were met in the application process to ensuring that FTA's requirements were met. Since FEMA's and FTA's programs have different statutory requirements, the agencies are limited in creating standardized procedures that would meet the requirements of both programs. Thus, transit agencies may be uncertain which program requirements they must meet until recovery efforts are well under way, as was the case with Hurricane Sandy.

FTA and FEMA Could Take Additional Steps to Help Ensure Effective Collaboration

Because FTA and FEMA face challenges assuming definitive and stable roles and responsibilities, it is especially important that they collaborate effectively. Among the practices that we have identified that can enhance and sustain the collaborative efforts of federal agencies is creating a means to monitor, evaluate, and report the results of collaborative efforts to better identify areas for improvement. Reporting on these activities can help key decision makers within the agencies, as well as clients and stakeholders, to obtain feedback for improving both policy and operation effectiveness.[50] We have identified implementation approaches among agencies that successfully monitored progress including developing performance measures and tying them to shared outcomes, identifying and sharing relevant agency performance data, developing methods to report on the group's progress that are open and transparent, and incorporating interagency group activities into individual performance expectations.[51]

FTA and FEMA have not implemented a method to monitor or evaluate their collaborative efforts or report on how well they are working. Instead, FTA and FEMA officials have noted that they meet regularly and notify each other of all grant awards to prevent duplication of DRAA funding. However, we have reported in the past that routine interagency meetings can be dutifully attended without communicating any substantive information. Thus, simply meeting does not guarantee that collaboration is effective.[52] Further, while FTA and FEMA officials state that they are collaborating now on the distribution of DRAA funds, it is unclear how the agencies might work together during future disasters. For example, the memorandum of agreement that MAP-21 required stated that "FTA and FEMA will work together to develop a communications protocol and implement an effective communications program to ensure all interested parties are aware of their respective responsibilities for emergency preparedness, response, and recovery related to public transportation matters." FTA and FEMA officials stated that their informal meetings to discuss DRAA funding serve as the communications protocol indentified in the memorandum of agreement. FTA officials told us that they do not have a communications protocol for anything beyond DRAA funding, because they do not have funding outside of DRAA. However, the requirement to develop the memorandum of agreement to improve coordination was intended to frame the relationship of FTA and FEMA going forward to address transportation emergency activities, not just one disaster. Further, without a means to monitor, evaluate, and report the results of collaborative efforts—including FTA's and FEMA's communication program and protocols— it is unclear how FTA and FEMA officials would know if their collaborative methods were effective. In a future disaster, informal meetings between FTA and FEMA may not be sufficient to ensure that effective collaboration is taking place, particularly if both agencies are funding public transportation activities.

CONCLUSION

The scope of activities eligible for transportation emergency relief has expanded over time to include funding resiliency efforts that repair and restore facilities beyond pre-disaster conditions. FTA's new Public Transportation

Emergency Relief Program was designed to be more flexible when funding resiliency compared to existing disaster-relief programs within FHWA and FEMA, although the scope of activities eligible for funding under those programs have expanded as well. Furthermore, the DRAA provided, through FTA, a significant investment in resiliency and protection against future disasters and funding well beyond (almost twice as much as) the repair and recovery needs of New York and New Jersey, as well as significant flexibility on how to provide those funds to areas affected by Hurricane Sandy. Going forward, as FTA begins awarding funding for resiliency projects through DRAA, the scope of these projects may exceed costs of projects that have previously been funded through historically established emergency relief programs in FHWA and FEMA. Resiliency funding can represent an efficient use of funds that avoids the need for repair and recovery from future disasters. We have reported on the need to build resiliency into infrastructure decision making and the challenges of doing so. However, expanding the scope of activities eligible for transportation emergency relief can pose challenges to providing long-term sustainable funding for disasters. As we have previously reported, the amount of emergency relief funding available is limited in an austere budget environment and may not be sustainable if federal disaster costs continue to increase. As the scope of eligible activities for emergency relief funding continues to expand, additional demands are placed on emergency relief funding further exacerbating the problem of longer-term sustainability. Determining the appropriate limits on emergency relief eligibility criteria will involve tradeoffs and require weighing competing demands in the current budget environment. We have made recommendations in the past on this matter and are therefore not making new recommendations in this report.

Looking beyond Hurricane Sandy relief, FTA's Public Transportation Emergency Relief Program is inherently limited by its inability to fund any activities—recovery or resiliency—without specific congressional action. This limitation has two notable effects. First, FTA cannot immediately respond to disasters involving transit when they occur, such as providing a "quick release" of funds to transit agencies, as was the case with Hurricane Katrina. Second, FTA and FEMA are limited in their ability to clearly delineate the responsibilities and costs each agency will assume during future disasters. As a result, FTA's and FEMA's programs will continue to have the authority and potential to fund the same transit activities, fostering confusion and difficulties for applicants seeking federal assistance in the days after a disaster. This confusion could be exacerbated if a future transit disaster occurred and FTA funding was not sufficient to address recovery needs without FEMA funds and transit agencies had to go to both agencies for the same disaster. Over time, the decision to provide funding authority to any particular agency has proved difficult and the broader challenges of funding disasters involves tradeoffs and requires weighing competing demands. In the meantime, FTA and FEMA can take steps to ensure that within the scope of activities they can control, their collaborative efforts are working. Although the agencies have a memorandum of agreement, FTA and FEMA have not developed any means to monitor, evaluate or report on their collaboration. Furthermore, should FTA receive regular, ongoing funding sometime in the future, then it and FEMA could take action to more clearly delineate roles and responsibilities and identify the costs each agency should assume.

Recommendation for Executive Action

We recommend that the Secretary of Transportation and the Secretary of Homeland Security direct the Administrators of FTA and FEMA to establish specific guidelines to monitor, evaluate, and report the results of collaborative efforts—including their communications program and protocol— for Hurricane Sandy as well as future disasters.

Agency Comments

We provided a draft of this report to DOT and DHS for its review and comment. DHS agreed with our recommendation and DOT took no position. DHS and DOT each provided technical comments, which we incorporated as appropriate.

Sincerely yours,
David Wise
Director, Physical Infrastructure Issues

Appendix I: Objectives, Scope, and Methodology

This report examines 1) the progress the Department of Transportation (DOT) has made allocating, obligating, and disbursing the Disaster Relief

Appropriations Act , 2013 (DRAA) surface transportation funds; 2) how the Federal Transit Administration's (FTA) new emergency relief program compares to the Federal Emergency Management Agency's (FEMA) and Federal Highway Administration's (FHWA) emergency relief programs; and 3) the extent to which FTA and FEMA have implemented their memorandum of agreement to coordinate their roles and responsibilities when providing assistance to transit agencies affected by Hurricane Sandy or future disasters.

To address all three objectives, we reviewed and analyzed relevant laws and regulations and our previous reports. We interviewed headquarters officials responsible for emergency relief programs and budgeting at FEMA, FTA, FHWA, and the Federal Railroad Administration (FRA). We also interviewed FTA regional officials responsible for emergency transit relief in New York and New Jersey, spoke with New York state officials and obtained background on the effects of Hurricane Sandy, and observed residual damage to transit facilities in New York and New Jersey. We interviewed officials from seven of the 13 local transit agencies that received grants from FTA.[1] We selected the transit agencies that received FTA's four largest funding allocations (all located within New York and New Jersey), as well as three agencies that received smaller funding allocations outside of New York and New Jersey. While our observations of these agencies provide key insights and illustrate FTA's emergency transit relief grant process, the results of our review should not be used to make generalizations about all FTA emergency transit relief grants. We also interviewed Amtrak—the sole grantee that received emergency relief grants from FRA (funds that had been transferred from FTA).[2] We did not interview FHWA grantees, as they did not receive any transferred FTA funds.

To determine the progress DOT has made allocating, obligating, and disbursing DRAA surface transportation funds from March 2013 to April 2014, we obtained and analyzed the financial information from DOT's surface transportation agencies (FTA, FHWA, and FRA) on Hurricane Sandy relief allocations, obligations, and disbursements. We also reviewed the FTA's Hurricane Sandy damage assessment of public transit agencies and the 2012 Congressional Budget Office analysis of the DRAA as well as FTA's announcements of Hurricane Sandy disaster relief availability of funds, interim rules, and allocations of funds published in the Federal Register. To access the reliability of the data we reviewed relevant documentation, interviewed knowledgeable officials, and reviewed internal controls. We found the data to be sufficiently reliable for our purposes.

To determine how FTA's new emergency relief program compares to FEMA's and FHWA's emergency relief programs, we interviewed headquarters officials responsible for emergency relief programs at FTA, FHWA, and FEMA.[3] In addition, we reviewed and analyzed applicable authorizing legislation for emergency relief programs for FEMA,[4] FTA,[5] and FHWA.[6] We also reviewed and analyzed emergency relief program regulations, procedures, guidance, and other agency documents on emergency relief at FEMA, FTA, and FHWA to determine similarities and differences between the programs' eligibility rules and requirements (see table 4). Finally, we reviewed independent reports on budgeting for emergency relief,[7] and our past reports on emergency relief efforts.[8]

Table 4. Agency Guidance Reviewed

Source	Guidance reviewed
Department of Homeland Security, FEMA	
	FEMA 322 Public Assistance Guide, June 2007 FEMA 323 Applicant Handbook, March 2010 FEMA 321 Public Assistance Policy Digest, January 2008 FY 2013 508 Hazard Mitigation Assistance Guidance Unified Guidance, (July 12, 2013) Final Addendum to the Hazard Mitigation Assistance Unified Guidance, (July 12, 2013)
Department of Transportation, FTA	
	Notice of Availability of FTA Emergency Relief Funds in Response to Hurricane Sandy, 78Federal Register 8691 (Feb. 6, 2013). Notice of Allocation of Public Transportation Emergency Relief Funds in Response to Hurricane Sandy, 78 Fed. Reg. 19357 (March 29, 2013). Issuance of Interim Final Rule, FTA Emergency Relief Program, 78 Federal Register 19136 (March 29, 2013). Notice of Second Allocation of Public Transportation Emergency Relief Funds in Response to Hurricane Sandy: Response, Recovery and Resiliency, 78 Federal Register 32296 (May 29, 2013). Transit Emergency Relief Program, Disaster Relief

Source	Guidance reviewed
	Appropriations Act of 2013, Grant Making and Grant Management Toolkit, (Apr. 12, 2013). Notice of Funding Availability for Resilience Projects in Response to Hurricane Sandy, 78 Fed. Reg. 78486 (Dec. 26, 2013).
Department of Transportation, FHWA	
	Emergency Relief Manual (Federal-Aid Highways),(May 31, 2013). A Guide to the Federal-Aid Highway Emergency Relief Program,(Nov. 5, 2012).

Source: GAO.

To determine the extent to which FTA and FEMA have implemented their memorandum of agreement to coordinate their roles and responsibilities when providing assistance to transit agencies affected by Hurricane Sandy or future disasters, we reviewed and analyzed the mandated memorandum of agreement between FTA and FEMA dated March 4, 2013 and compared the agencies' efforts to practices that we have found can enhance collaboration among federal agencies.[9] We evaluated the extent to which the memorandum of agreement defined roles and responsibilities, assigned costs, specified a system for monitoring and evaluating collaborative efforts, and established communication protocols. We developed a list of evaluation questions based upon the requirements of FTA's Public Transportation Emergency Relief Program authorization in MAP-21 and funding requirements in DRAA. Three GAO team members independently analyzed the memorandum of agreement. The analyses were then consolidated and agreed to by the team members and reviewed by the GAO team's Assistant Director. We also reviewed the DOT OIG report on FTA's oversight of emergency relief funds for coordination between FTA and FEMA for transit relief grants.[10]

We conducted this performance audit from April 2013 to May 2014 in accordance with generally accepted government auditing standards. Those standards require that we plan and perform the audit to obtain sufficient, appropriate evidence to provide a reasonable basis for our findings and conclusions based on our audit objectives. We believe that the evidence obtained provides a reasonable basis for our findings and conclusions based on our audit objectives.

End Notes

[1] The Robert T. Stafford Disaster Relief and Emergency Assistance Act, as amended April 2013, establishes the authority under which states request a presidential disaster declaration. The act also established the programs and process through which the federal government provides disaster assistance to state and local governments, tribes, and certain nonprofit organizations and individuals. 42 U.S.C. §§ 5121-5207.

[2] Pub. L. No. 113-2 div. A, 127 Stat. 4 (Jan. 29, 2013).

[3] The Moving Ahead for Progress in the 21st Century Act (MAP-21) requires that the Secretaries of the Departments of Homeland Security and Transportation enter a memorandum of agreement to improve coordination between the two departments in providing assistance for public transportation and expedite the provision of federal assistance relating to a major disaster or emergency declared by the President under the Stafford Act. Pub. L. No. 112-141, § 20017(b), 126 Stat. 405, 705 (July 6, 2012). FTA and the Federal Emergency Management Agency (FEMA) entered into the memorandum of agreement on behalf of DOT and DHS respectively.

[4] The National Railroad Passenger Corporation (Amtrak).

[5] During the review the team held coordination meetings with the DOT OIG. We learned they are conducting an extensive multi-phase review of how FTA's Public Transportation Emergency Relief program is being set up and FTA's efforts to ensure that DRAA funds are distributed appropriately. Thus, the team made a decision to focus our review on how the program is structured compared to previously established emergency relief programs in FEMA and FHWA instead of conducting an in-depth evaluation of FTA's Public Transportation Emergency Relief Program at this time.

[6] According to CBO, this law increased nondefense discretionary appropriations for fiscal year 2013 by more than $50 billion and defense discretionary appropriations by over $100 million.

[7] Pub. L. No. 112-141, § 20017(a), 126 Stat 405, 703, codified at 49 U.S.C. § 5324.

[8] FTA's DRAA appropriation has "no-year authority." No-year authority or no-year funding refers to appropriations that do not restrict the time by which funds must be obligated. The DRAA requires that funds for surface transportation grants be expended by grantees within 24 months following the obligation of funds for the grant, unless the Director of the Office of Management and Budget (OMB) waives the requirement and notifies the congressional Committees on Appropriations. On July 9, 2013, OMB waived this requirement for FTA, FRA, and FHWA.

[9] 49 C.F.R. § 602.5.

[10] These projects must be authorized under titles 23 or 49 of the United States Code, which primarily pertain to highways and bridges, and rail and transit, respectively.

[11] The absence of legislation to reduce the federal budget deficit by at least $1.2 trillion by fiscal year 2021 triggered the sequestration process in section 251A of the Balanced Budget and Emergency Deficit Control Act of 1985 as amended. On March 1, 2013, the President ordered a sequestration of $85.3 billion in discretionary appropriations and direct spending for fiscal year 2013. OMB calculated that this would amount to a 5 percent reduction in nonexempt nondefense discretionary funding for the fiscal year.

[12] For purposes of this report, "allocated" reflects the amount of available funds that have been assigned by DOT to a grantee from which grants may be awarded. "Obligated" reflects the amount of funds legally committed by DOT to a grantee for awarded grants. "Disbursed" reflects the amount of payments that have been made to a grantee.

[13] In March 2013, FTA allocated $2 billion (less almost $21 million to fund expenses for FTA administration and oversight by the DOT OIG) to affected recipients for eligible emergency response and recovery costs including (1) $576.6 million to reimburse affected transit agencies for emergency response and recovery costs incurred or budgeted to date and (2) $1.4 billion primarily to the four agencies most severely impacted (see table 1) for repair and reconstruction projects. These reimbursements are proportional to each agency's estimated overall recovery costs. The $1.4 billion includes the FTA set-aside of approximately $28 million for other affected agencies that may have additional response and recovery expenses not reimbursed to date. In May 2013, FTA allocated an additional $3.7 billion to the four agencies most severely affected, based on a percentage of the anticipated full cost of recovery and rebuilding.

[14] Congressional Budget Office, Estimate of the Disaster Relief Appropriations Act, 2013, as Posted on the website of the Senate Committee on Appropriations on December 12, 2012 (Washington, D.C.: Dec. 12, 2012).

[15] 78 Fed. Reg. 78486 (Dec.26, 2013).

[16] These funds are available for states, local governmental authorities, tribal governments, and other FTA recipients within the declared Hurricane Sandy disaster area. Entities that provide public transportation service and are not current recipients of FTA funding are only eligible to receive emergency relief funding as a subrecipient of an FTA recipient.

[17] This is defined as areas for which the President declared a major disaster under the Stafford Act in response to Hurricane Sandy.

[18] The Notice of Funds Availability defines resiliency projects as those projects designed and built to address current and future vulnerabilities to a public transportation facility or system due to future occurrence or recurrence of emergencies or major disasters that are likely to occur in the geographic area in which the public transportation is located; or projected changes in development patterns, demographics, or climate change and extreme weather patterns.

[19] In December 2012, the President signed an executive order creating the Hurricane Sandy Rebuilding Task Force to ensure that the federal government continues to provide appropriate resources to support affected State, local, and tribal communities to improve the region's resilience, health, and prosperity by building for the future. 77 Fed. Reg. 74341 (Dec. 14, 2012). The Task Force is chaired by the Secretary of Housing and Urban Development and is comprised of representatives from other federal agencies.

[20] Hurricane Sandy Rebuilding Task Force, Hurricane Sandy Rebuilding Strategy: Stronger Communities, A Resilient Region (Washington, D. C.: August 2013).

[21] Department of Transportation, Office of Inspector General, MH-2014-008, Federal Transit Administration: Initial Assessment of FTA's Oversight of the Emergency Relief Program and Hurricane Sandy Relief Funds (Washington, D. C.: Dec. 3, 2013).

[22] The amount of Hazard Mitigation Grant Program funding available to an applicant is based upon the estimated total of Federal assistance, subject to the sliding scale formula outlined in section 5170c(a) of title 42, that FEMA provides for disaster recovery under the Presidential major disaster declaration. The formula provides for up to 15 percent of the first $2 billion of estimated aggregate amounts of disaster assistance, up to 10 percent for amounts between $2 billion and $10 billion, and up to 7.5 percent for amounts between $10 billion and $35.333 billion. For States with enhanced plans—FEMA approved state plans identifying policies and actions that can be implemented over the long term to reduce risks and future losses—the eligible assistance is up to 20 percent for estimated aggregate amounts of disaster assistance not to exceed $35.333 billion.

[23] FEMA may provide discretionary hazard mitigation funding under Section 406 of the Stafford Act which refers to FEMA's Public Assistance Program. Section 406 is applied to the parts of the facility that were actually damaged by the disaster and the mitigation measure provides protection from subsequent events. The mitigation work must be cost effective and be reasonably performed as part of the work or measure that will reduce the potential for damage to a facility from a disaster event. In addition, Community Development Block Grant Program (CDBG) funds—including Community Development Block Grant Disaster Recovery (CDBG–DR) funds may be used for the payment of the non-federal share required in connection with a federal grant-in-aid program (e.g., the FEMA Public Assistance Program) that provides funding for a CDBG-eligible activity.

[24] 49 C.F.R.§ 602.13(C)(5).

[25] 23 U.S.C. § 120(i)(1)(A).

[26] For some disasters, non-cash contributions referred to as "in-kind donations" may be credited toward the non-federal share of grant costs under FEMA's Public Assistance and Hazard Mitigation programs. In order for an in-kind contribution to be eligible for the nonfederal share, several conditions must apply such as that the contribution be necessary to accomplish the project objectives. Examples of in-kind contributions include property, services, equipment, and donated materials.

[27] GAO, Emergency Transit Assistance: Federal Funding for Recent Disasters, and Options for the Future, GAO-08-243, (Washington D.C.: Feb. 15, 2008) and GAO, Disaster Recovery: FEMA's Public Assistance Grant Program Experienced Challengers with Gulf Coast Rebuilding, GAO-09-129 (Washington, D. C.: Dec. 18, 2008).

[28] The total cost of a project funded through FHWA's Emergency Relief Program may not exceed the cost of repairing or constructing a facility that meets the current geometric and construction standards required for the types and volume of traffic that the facility will carry over its design life. Pub. L. No. 112-141, § 1107, codified at 23 U.S.C. § 125(d)(2).

[29] Pub. L. No. 113-2, div. B, 127 Stat. 4, 39-50.

[30] GAO, Highway Emergency Relief: Reexamination Needed to Address Fiscal Imbalance and Long-term Sustainability, GAO-07-245 (Washington D.C.: Feb. 23, 2007).

[31] GAO-08-243.

[32] GAO, Climate Change: Future Federal Adaptation Efforts Could Better Support Local Infrastructure Decision Makers, GAO-13-242 (Washington, D.C.: Apr. 12, 2013).

[33] GAO-13-242.

[34] FEMA has several Hazard Mitigation Assistance Grant Programs that provide funding for pre- and post-disaster mitigation. Some of FEMA's other hazard mitigation programs are not funded by the DRF. These programs include the Pre-disaster Mitigation Program and Flood Mitigation Assistance.

[35] For example, the President's Budget Request for the DRF for fiscal year 2009 was based on the 5-year average obligational level for non-catastrophic disaster activity (less estimates of funds carried forward and the amount of funds deobligated in prior years that can be used for new obligations for authorized purposes).

[36] Contract authority is the statutory authority to incur obligations in advance of an appropriation. Contract authority is unfunded, and a subsequent appropriation is needed to liquidate the obligation. FHWA obligates funds for a state project once FHWA approves a project. Subsequently, FHWA reimburses states out of the Highway Account of the Highway Trust Fund ("Highway Account") or the General Fund, depending on the source of funding. Congress can and has provided FHWA with supplemental appropriations, such as the DRAA, which are not funded through the Highway Trust Fund. The FHWA Emergency

Relief Program is a reimbursable program—meaning FHWA reimburses the state for the federal share of the cost of the completed work assuming all federal requirements are met.

[37] 23 U.S.C. § 125.

[38] GAO-08-243.

[39] We also presented other options for Congress to consider such as 1) making permanent the temporary authorities that FTA was given to respond to the 2005 Gulf Coast hurricanes and (2) incorporating transit assistance as an eligible activity within FHWA's Emergency Relief Program. We also found that at the local and state level, mutual aid agreements between states and others can direct needed resources to transit agencies following a disaster.

[40] FEMA's Public Assistance Program does have an expedited payment process for debris removal and emergency protective measures. Under this process, FEMA will make payments for these activities within 60 days of completion of the estimated preliminary damage assessment and no later than 90 days after the pre-application was submitted. However, this funding is only available on a project-by-project basis, and FEMA is only able to release less than $1 million. Funding beyond $1 million would require OMB and congressional approval.

[41] GAO, High Risk Series: An Update, GAO-13-283 (Washington. D.C.: February 2013).

[42] GAO, Federal Disaster Assistance: Improved Criteria Needed to Assess a Jurisdiction's Capability to Respond and Recover on its Own, GAO-12-838 (Washington, D.C.: Sept. 12, 2012).

[43] GAO-07-245. In order to help put the FHWA Emergency Relief Program on a sustainable and more sound financial footing, we recommended among other things, that FHWA should, within its authority, tighten eligibility standards, recapture unused funds, and seek rescission of unneeded funds.

[44] GAO, Supplemental Appropriations: Opportunities Exist to Increase Transparency and Provide Additional Controls, GAO-08-314, (Washington D.C.: Jan. 31, 2008).

[45] Erskine Bowles, Alan Simpson et al., The Moment of Truth, Report of the National Commission on Fiscal Responsibility and Reform (December 2010).

[46] CRS, Disaster Relief Funding and Supplemental Appropriations for Disaster Relief, CRS Report, R40708 (Washington D.C.: Aug. 5, 2013).

[47] GAO, Results-Oriented Government: Practices That Can Help Enhance and Sustain Collaboration among Federal Agencies, GAO-06-15 (Washington, D.C.: October 2005).

[48] 49 U.S.C.§ 5324(c)(2).

[49] The President made major disaster declarations in the states of New Hampshire on March 19, 2013; Connecticut on March 21, 2013; Rhode Island on March 22, 2013; Maine on March 25, 2013; Massachusetts on April 19, 2013; and New York on April 23, 2013.

[50] GAO-06-15.

[51] GAO, Managing for Results: Implementation Approaches Used to Enhance Collaboration in Interagency Groups , GAO-14-220 (Washington, D.C.: Feb. 14, 2014).

[52] GAO-06-15.

End Notes for Appendix I

[1] New Jersey Transit Corporation, Port Authority of New York and New Jersey, New York City Department of Transportation, New York Metropolitan Transportation Authority,

Massachusetts Bay Transportation Authority, Southeastern Pennsylvania Transportation Authority, and Milford Connecticut Transit District.

[2] The National Railroad Passenger Corporation (Amtrak).

[3] During the review the team held coordination meetings with the DOT OIG. We learned they are conducting an extensive multi-phase review of how the FTA's Public Transportation Emergency Relief Program is being set up and FTA's efforts to ensure that DRAA funds are distributed appropriately. Thus, the team made a decision to focus our review on how the program is structured compared to previously established emergency relief programs in FEMA and FHWA instead of conducting an in-depth evaluation of FTA's Public Transportation Emergency Relief Program at this time.

[4] 42 U.S.C. §§ 5170, 5170c, 5172; 44 C.F.R. Part 206.

[5] 49 U.S.C. § 5324; 49 C.F.R. Part 602.

[6] 23 U.S.C. §125 ; 23 C.F.R. Part 668.

[7] Congressional Research Service, Disaster Relief Funding and Supplemental Appropriations for Disaster Relief, August 5, 2013 and Erskine Bowles, Alan Simpson et al., The Moment of Truth, Report of the National Commission on Fiscal Responsibility and Reform (December 2010).

[8] GAO-07-245 and GAO-08-243.

[9] GAO-06-15.

[10] Department of Transportation, Office of Inspector General Audit Report, Initial Assessment of FTA's Oversight of the Emergency Relief Program and Hurricane Sandy Relief Funds, Federal Transit Administration, Report Number: MH-2104-008, (Washington, DC, Dec. 3, 2013).

INDEX

A

B

C

D

E

F

G

H

I

L

M

N

O

P

T

U

V

W